BIBLIOGRAPHY of the Arctic Arthropods of the Nearctic Region

Compiled by H.V. Danks
Biological Survey Project
Entomological Society of Canada
and
Biological Survey of Canada
(Terrestrial Arthropods)
National Museum of Natural Sciences
Ottawa

Published by the
ENTOMOLOGICAL SOCIETY OF CANADA
Ottawa K1Z 7K9
1981

ISBN 0-9690829-1-6

CONTENTS

PREFACE

This bibliography, a product of the Biological Survey Project of the Entomological Society of Canada, results from a 1978-1980 contract between the Society and the Government of Canada for a Review and synthesis of knowledge on northern and arctic insects. The contract was funded by the National Museum of Natural Sciences, which acted as lead department, by Supply and Services Canada, and the Departments of National Defence, Indian and Northern Affairs, and Energy, Mines and Resources (Geological Survey of Canada).

In addition, library facilities were kindly made available by the National Museum of Natural Sciences, the Geological Survey of Canada, Agriculture Canada, the National Research Council (Canada Institute for Scientific and Technical Information), and by the Arctic Institute of North America in Calgary and the Boreal Institute for Northern Studies in Edmonton.

INTRODUCTION

Coverage

This bibliography deals with the terrestrial arthropods of arctic North America and Greenland, including only animals there that breathe air at some stage of the life cycle (Insecta, Collembola, Arachnida, Chilopoda, Protura). Freshwater crustaceans are thereby excluded.

The area covered includes the whole of Greenland, and the tundra north of tree line in North America, but not tundra areas of southwestern Alaska and the Aleutian Islands and southern alpine tundras. Danks (1981) discussed the reasons for adopting this definition of the North American arctic.

Subarctic areas (localities within the open boreal forest or forest-tundra) are excluded. Churchill, Manitoba - a site at tree line - is however specifically included because, following the collection of 50,000 invertebrate specimens there by McClure (1943, p. 34), much other material was collected during the Canadian Northern Insect Survey (cf. Freeman 1959). The Pribilof Islands, although in the maritime subarctic zone, are also included, as they are isolated in the ocean and have an impoverished fauna that has been relatively well collected.

Citations are included even if they contain information only for a single arctic locality, and for a single species treated in a longer work. Some interim reports containing little data have not been cited if duly published summaries of the research now exist and are cited here.

Previous arctic bibliographies

The most useful single source for the literature of the arctic regions is the *Arctic Bibliography* (Arctic Institute of North America) (1953-1975). Subsequent literature is covered by a variety of computer-based abstracting services, but their coverage is less effective for northern regions than was the coverage by Arctic Bibliography (Orr 1978). The library catalogue of the Scott Polar Research Institute (e.g. 1976) is a particularly useful list of arctic references, and is supplemented by subsequent issues of *Recent Polar Literature* (formerly part of *Polar Record*). Recent accessions to the library of the Arctic Institute of North America are listed by the ASTIS (Arctic Science and Technology Information System) Current Awareness Bulletin. For orientation to older sources see Dutilly (1945). Entomologists should also be aware of the useful general introduction to the literature compiled by Trauger *et al.* (1974).

Other bibliographies of such topics as fisheries (Nicholson 1974-1976) or northern ecology (Peterson 1974-1978) deal partly with insects and their relatives. There are many other more specialized bibliographies (e.g. (e.g. Roberts-Pichette 1972), some of them dealing exclusively with particular insects, especially biting flies (e.g. Dutilly 1950; Mezenev 1971).

Several regional bibliographies exist, as for the Yukon (Lotz 1964; Hemstock and Cooke 1973, 1975; Cooke *et al.* 1977; Thomas *et al.* 1978), Alaska (Tourville 1974), the Mackenzie Delta (Jones 1969), northern Ontario (Spencer and Holland 1968), and northern Manitoba (Teillet 1979), but coverage for arthropods is generally incomplete and confined to subarctic areas. Nettleship and Smith (1975) cited many papers from arctic Canada.

Some northern research institutions have published bibliographies of their work. Gunn's (1973) list is especially relevant here. Organizations, especially government departments, have also contributed bibliographies (e.g. Davies 1969; Hobson and Voyce 1974-1980).

Methods of Compilation

This bibliography was prepared from citations secured primarily:

(1) By a manual search of all volumes of Arctic Bibliography, through the index.

(2) By a computer search of Biological Abstracts (BIOSIS) 1969-mid 1980.

(3) By inspection of the reference lists of papers traced by these means.

Some additional information was obtained from the other sources cited in the previous section, and from the files of the compiler. Coverage is therefore essentially comprehensive through mid-1979.

Relevant papers were inspected and information was extracted on the general content and on the numbers of references, illustrations, tables, and keys if any. Key words were noted for later use in preparing the geographic, taxonomic, and subject indexes.

Some general difficulties with citations

Papers have sometimes been published in more than one place, in more than one format, or in different years, especially in older work. For instance, Elwes (1903) paper was republished as a chapter of Hanbury's (1904) book. Citation is then possible by chapter author or by book author. Both citations are given here in such cases, but only the more specific citation has been indexed.

Identical items are sometimes published in two formats, even today. For example, the journal Rep. Kevo Subarctic Res. Stn is identical in content and pagination to some volumes of the Ann. Univ. Turku, Series A.2 Biol-geogr.; only the colour of the cover, and sometimes the year of publication, differ.

Some journals consist of selected papers compiled from scattered sources, with or without repagination. Early issues of the Norsk Polarinstitutt Årbok were compiled in this way. Other compilations are more easily noticed from the title, as Oxford Univ. Exploration Club Greenland and Spitsbergen Papers.

Another difficulty arises from journal title changes. Authors may cite, and libraries may even bind, the early issues under the newer name, and thus for instance Norges Svalbard- og Ishavs- Undersøkelser Meddr becomes [incorrectly] cited as Norsk Polarinstitutt Meddr.

Other journals are numbered serially in volumes, but the parts of each volume may appear in different years (e.g. Meddr Grønland), sometimes leading to dating errors in subsequent author's citations.

Notes to explain these and other complications are included in this bibliography when known, but users should be alert for other examples.

A second group of difficulties arises from the existence of different transliterations from languages such as Russian (e.g. Kurentsov or Kurenzov). However, there are few papers in Russian on North American insects, though the same Russian author's name may have been published in different forms if papers by the same author in Romance or Scandinavian languages have appeared in different journals. Note also that the vowels ä, å, ö, ø and ü, are sometimes rendered ae, aa, oe, oe and ue, even in the original journals.

References for the introduction

Cooke, G.A., G.D. Perry, and M.F. Ridge. 1977. Yukon bibliography. Update to 1975. Occ. Pub. Boreal Inst. N.Stud., Univ. Alta. 8-3. 408 pp.

Danks, H.V. 1981. Arctic Arthropods. A review of systematics and ecology with particular reference to the North American fauna. Entomological Society of Canada, Ottawa. 608 pp.

Davies, G.K. 1969. An annotated bibliography of unclassified reports issued by Defence Research Northern Laboratory 1947-1965. Defence Scientific Information Service (DSIS) Rep. No. B-13. Defence Res. Bd Can., Ottawa. 68 pp.

Dutilly, A. 1945. Bibliography of bibliographies on the arctic. Catholic Univ. of America, Washington. 47 pp. (Catholic Univ. Am. Publ. 1B)

Dutilly, A. 1950. A bibliography on the relation of mosquitoes to vegetation in the Eurasian Arctic and Subarctic. U.S. Army. Office of Quartermaster General. Environmental Protection Section. Rep. 162. Contract W44-109-qm-1297.

Elwes, H.J. 1903. On a collection of Lepidoptera from Arctic America. Trans. ent. Soc. Lond. 1903(3): 239-243.

Freeman, T.N. 1959. The Canadian Northern Insect Survey, 1947-1957. Polar Rec. 9(61): 299-307.

Gunn, W.W. 1973. Bibliography of the Naval Arctic research laboratory. Arct. Inst. N. Am. Tech. Pap. 24. 176 pp.

Hanbury, D.T. 1904. Sport and travel in the northland of Canada. MacMillan Co., N.Y. 319 pp.

Hemstock, C.A. and G.A. Cooke. 1973. Yukon bibliography. Update 1963-1970. Occ. Pub. Boreal Inst. N.Stud., Univ. Alta. 8-1. 420 pp.

Hemstock, C.A. and G.A. Cooke. 1975. Yukon bibliography. Update to 1973 Occ. Pub. Boreal Inst. N.Stud., Univ. Alta. 8-2. 384 pp.

Hobson, G.D. and J. Voyce (compilers). 1974-1980. Polar Continental Shelf Project. Titles and abstracts of scientific papers supported by PCSP. No. 1 (1974), 76 pp.; no. 2 (1975), 68 pp.; no. 3 (1977), 97 pp.; no. 4 (1980), 89 pp.

Jones, M.J. (Ed.). 1969. Mackenzie Delta Bibliography. Can. Dep. Indian Affairs N. Dev., N. Sci. Res. Group, Ottawa. 119 pp.

Lotz, J.R. 1964. Yukon bibliography. Yukon Res. Proj. Rep. 1.

McClure, H.E. 1943. Aspection in the biotic communities of the Churchill area, Manitoba. Ecol. Monogr. 13(1): 1-35.

Mezenev, N.P. 1971. [The distribution of mosquitoes in the northern arctic regions. (Survey of literature on the subject)]. Trudy Nauchno-Issledovatel'skiy Inst. Sel'skogo Khozyaystva Kraynego Severa 19: 191-207. [In Russian]

Nettleship, D.N. and P.A. Smith (Eds.). 1975. Ecological sites in northern Canada. Can. Comm. for I.B.P., Conservation Terrestrial - Panel 9. 330 pp.

Nicholson, H.F. 1974-1976. Bibliography on the limnology and fisheries of Canadian freshwaters. Can. Fish. Marine Serv. Tech. Rep. 471, 1974(1); 504, 1975(2); 566, 1975(3);, 672, 1976(4); 605, 1976 (Appendix A, cumulative author index).

Orr, J.R. 1978. Study of the Arctic Bibliography and the abstracting and indexing services. Arct. Bull. 2(12): 295-302.

Peterson, N.M. (compiler) 1974-1978. North of 60. Ecology of the Canadian Arctic Archipelago: selected references. Vol. 1(1974), pp. 1-425; Vol. 2 (1974), pp. 426-846; Vol. 3 (1975), pp. 847-1304; Vol. 4(1976), pp. 1305-1790; Vol. 5 (1978), pp. 1791-2277. Indian and Northern Affairs/Information Canada.

Roberts-Pichette, P. 1972. Annotated bibliography of permafrost-vegetation-wildlife-landform relationships. Forest Management Institute, Information Report FMR-X-43. Can. Forestry Serv., Ottawa. 350 pp.

Scott Polar Research Institute. 1976. The Library Catalogue of the Scott Polar Research Institute, Cambridge, England. [incl. subject catalogue (esp. vol. 12, 592-599)] G.K. Hall, Boston, Mass.

Spencer, L. and S. Holland (compilers). 1968. Northern Ontario, a bibliography. Univ. Toronto Press, Toronto. 120 pp.

Teillet, D.J. 1979. A northern Manitoba bibliography. Tech. Rep. Man. Dep. Mines, Nat. Res., Environ. 79-5.

Thomas, H.L., G.A. Cooke, and G.D. Perry. 1978. Yukon bibliography. Update to 1977. Occ. Pub. Boreal Inst. N. Stud., Univ. Alta. 8-4. 476 pp.

Tourville, E.A. 1974. Alaska, a bibliography 1570-1970, with subject index. G.K. Hall, Boston, Mass. 738 pp.

Trauger, S.C., R.D. Shenefelt, and R.H. Foote. 1974. Searching entomological literature. Bull. ent. Soc. Am. 20(4): 303-315.

HOW TO USE THIS BIBLIOGRAPHY

Indexes

Four methods of access to the information in titles listed here are provided.

(1) Alphabetical access by author (first-named authors in the body of the bibliography, subsequent authors in a separate index). Entomological sections within more general works are cross referenced from the general author or editor (compare Murdoch and Riley 1885, #947; Ray *et al*. 1885, #1045).

(2) Access by area. Three main regions are treated here: Canada, Alaska and Greenland. Within these regions, only localities cited relatively frequently are indexed, and only for papers that deal with one or a few such sites. Other papers are indexed by larger regions. Greenland is normally indexed by region (E, W, etc.).

(3) Access by taxon. Arthropod taxa are indexed by family if only a few (<4) families are treated in a paper, or by order or comparable subdivision if only a few (<4) orders are treated, or by class. Papers dealing only with taxonomy, descriptions, and records of distribution are distinguished from those with different or wider focus.

(4) Access by subject. The subject index contains appropriate subject cross-references, and deals chiefly with ecological and zoogeographical aspects. Purely taxonomic papers are indexed only in (3).

Abbreviations

Abbreviations of journal titles follow the latest edition of the *World List of Scientific Periodicals*.

The only other abbreviations used in the main part of the bibliography are ref.(s) [reference(s)], fig.(s) [figure(s)], pl.(s) [plate(s)], and tab.(s) [table(s)]. Numbers of references, and of illustrations and tables if any are given in parentheses after each citation, except for references cited only in the text and unnumbered tables in the text.

LIST OF CITATIONS

1. Addison, J.A. 1975. Ecology of Collembola at a high arctic site, Devon Island, N.W.T. Unpub. Ph.D. thesis, Univ. Calgary. 212 pp.

2. Addison, J.A. 1977. Population dynamics and biology of Collembola on Truelove Lowland. pp. 363-382 in L.C. Bliss (Ed.), Truelove Lowland, Devon Island, Canada: A high arctic ecosystem. Univ. Alta Press, Edmonton. 714 pp. (34 refs., 11 figs., 8 tabs.)

3. Addison, J.A. and D. Parkinson. 1978. Influence of collembolan feeding activities on soil metabolism at a high arctic site. Oikos 30(3): 529-538. (51 refs., 3 figs., 4 tabs.)

4. Aldrich, J.M. 1918. Notes on Diptera. Psyche, Camb. 25(2): 30-35. (text refs.)

5. Aldrich, J.M. 1929. Revision of the two winged flies of the genus Coelopa Meigen in North America. Proc. U.S. natn. Mus. 76(11): 2808, 1-6. (text refs., key)

6. Aldrich, J.M. and P.S. Darlington. 1908. The dipterous family Helomyzidae. Trans. Am. ent. Soc. 34(1): 67-100. 2 pl. (text refs., 9 figs., keys)

7. Alexander, C.P. 1919. Part C: Diptera. The Crane-flies collected by the Canadian Arctic Expedition, 1913-1918. Rep. Can. Arctic Expedition 1913-1918. Vol. III: Insects.pp. 1-30. (text refs., 65 + 2 figs.)

8. Alexander, C.P. 1921. Dipterous insects of the family Tipulidae from the Pribilof Islands, Alaska. pp. 183-184 in G.D. Hanna et al. Insects of the Pribilof Islands, Alaska. Proc. Calif. Acad. Sci. ser. 4, 11(14): 153-195. (text ref.)

9. Alexander, C.P. 1923a. Diptera, suborder Orthorrhapha, division Nematocera, families Tipulidae and Rhyphidae. pp. 159-169 in W.L. McAtee (Ed.), A biological survey of the Pribilof Islands, Alaska. Part II, Insects, arachnids and chilopods of the Pribilof Islands Alaska. N. Am. Fauna 46: 129-255. (text refs., 23 figs.)

10. Alexander, C.P. 1923b. Report of the second Norwegian arctic expedition in the "Fram" 1898-1902. The Crane-Flies (Tipulidae, Diptera). Norsk ent. Tidsskr. 1: 296-297. (text refs.)

11. Alexander, C.P. 1934. II. Diptera collected on Southampton Island by G.M. Sutton. Trichoceridae and Tipulidae. The exploration of Southampton Island, Hudson Bay, by G.M. Sutton. Mem. Carnegie Mus. 12, part II, sect. 4: 3-9. (text + 29 refs., 10 figs.)

12. Alexander, C.P. 1938. New or insufficiently known crane flies from the nearctic region (Tipulidae, Diptera), part IV. Bull. Brooklyn ent. Soc. 33(2): 71-78. (text refs.)

13. Alexander, C.P. 1946. Undescribed species of crane-flies from the eastern United States and Canada (Dipt: Tipulidae). Part X. Ent. News 57(10): 245-252. (0 ref.)

14. Alexander, C.P. 1955. The crane flies of Alaska and the Canadian Northwest (Tipulidae, Diptera). The genus Erioptera Meigen.

Misc. Publs Mus. Zool. Univ. Mich. 90. 33 pp. (33 + text refs., 38 figs.)

15. Alexander, C.P. 1956. Two new crane flies from Point Barrow, Alaska. Pan-Pacif. Ent. 32(3): 123-125. (text refs.)

16. Alexander, C.P. 1973. Undescribed species of nearctic Tipulidae (Diptera), XI. Gt Basin Nat. 33(3): 189-196. (0 ref., 11 figs.)

17. Andersen, F.S. 1937. Ueber die Metamorphose der Ceratopogoniden und Chironomiden Nordost Grönlands. Meddr Grønland 116(1). 95 pp. (51 refs., 56 figs., 1 map, keys)

18. Andersen, F.S. 1946. East Greenland lakes as habitats for chironomid larvae. Studies on the systematics and biology of Chironomidae, II. Meddr Grønland 100(10). 65 pp. (36 refs., 1 fig., 2 pls., 20 tabs.)

19. Andrews, G.F. 1978. Taxonomy and host specificity of nearctic Alloxystinae with catalog of the world species (Hymenoptera: Cynipidae). Occ. Pap. Ent., St. Calif. Dep. Food Agric. 25. 128 pp. (114 refs., 114 figs., 8 tabs., keys)

20. Angus, R.B. 1970. A revision of the beetles of the genus Helophorus F. (Coleoptera: Hydrophilidae) subgenera Orphelophorus D'Orchymont, Gephelophorus Sharp and Meghelophorus Kuwert. Acta zool. fenn. 129. 56 pp. (65 refs., 2 figs. + 3 pls., keys)

21. Arctic Circular. 1948. Biting insects in the Arctic. Arct. Circ. 1: 67. (0 ref.)

22. Arndt, W. 1931. Die Copeognatha des arktischen Gebietes. Fauna arct. 6: 59-66. (20 refs.)

23. Arntfield, P.W. 1975. A revision of Graphomya Robineau-Desvoidy (Diptera: Muscidae) from North America. Can. Ent. 107(3): 257-302. (46 refs., 22 figs., 3 maps, key)

24. Asbirk, S. and N.E. Franzmann. 1977. Studies of snow buntings. pp. 132-142 in G.H. Green and J.J.D. Greenwood (Eds.), Report of the Joint Biological Expedition to N.E. Greenland 1974. Dundee: Dundee University N.E. Greenland Expedition. (refs., 7 figs., 7 tabs.)

25. Ashworth, A.C. 1977. A late Wisconsinan Coleopterous assemblage from southern Ontario and its environmental significance. Can. J. Earth Sci. 14(7): 1625-1634. (15 refs., 2 figs., 2 tabs.)

26. Ashworth, A.C. 1979. Quaternary Coleoptera studies in North America: past and present. pp. 395-405 in T.L. Erwin, G.E. Ball, and D.R. Whitehead (Eds.), Carabid beetles, their evolution, natural history and classification. W. Junk, The Hague. (39 refs., 2 figs., 1 appendix)

27. Attems, C.G. 1909. Die Myriopoden der Vega-Expedition. Ark. Zool. 5(3): 1-84. (text refs., 27 + 86 figs., 5 tabs.)

28. Atyeo, W.T. 1960. A revision of the mite family Bdellidae in North and Central America. Kans. Univ. Sci. Bull. 40(8): 345-499. (42 refs., 197 figs., keys)

29. Aurivillius, C. 1883. [Insect Life in Arctic Lands] pp. 403-459 in A.E. Nordenskiöld (Ed.), Studier Och Forskningar Föranledda af

Mina Resor I Hoga Norden. Beijers, Stockholm. 550 pp. (In Swedish). (text refs., text tabs.) [Translation: Boreal Institute, University of Alberta, Edmonton].

30. Aurivillius, C. 1890. Gronlands insektfauna. I. Lepidoptera, Hymenoptera. Svenska vetenskAkad. Handl. Bihang, ser. 4, 15 (1). 33 pp. (In Swedish) (27 refs., 4 figs., 3 pls.)

31. Aurivillius, C. 1900. [Lepidoptera and Coleoptera collected during Professor A.G. Nathorst's Arctic Expeditions, 1898 and 1899; during the Swedish Expedition to Bear Island 1899 and during curator G. Kolthoff's Expedition to Greenland 1900.] Ofvers. K. VetenskAkad. förh. 57: 1135-1144. (In Swedish) (0 ref.)

32. Baker, A.W. 1919. Part D: Mallophaga and Anoplura. Mallophaga of the Canadian Arctic Expedition, 1913-1918. Rep. Can. Arctic Expedition, 1913-1918. Vol. III: Insects. pp.3-9 (0 ref., 4 figs.)

33. Baker, C.F. 1904. A revision of American Siphonaptera, or fleas, together with a complete list and bibliography of the group. Proc. U.S. natn. Mus. 27(1361): 365-469. (text + 83 refs., 17 pls.)

34. Baker, M.C. 1977. Shore bird food habits in the eastern Canadian Arctic. Condor 79(1): 56-62. (14 refs., 4 figs., 5 tabs.)

35. Bakkendorf, O. 1955. Notes on Icelandic and Greenlandic Chalcidoideous Hymenoptera. Ent. Meddr 27(5): 135-162. (48 refs., 41 figs.)

36. Ball, G.E. 1962. Descriptions of eleven new species of the beetle subgenus Cryobius (Pterostichus, Col. Carabidae) from Alaska and northwestern Canada. Opusc. ent. 28(1-2): 1-26. (0 ref.)

37. Ball, G.E. 1963. The distribution of the species of the subgenus Cryobius (Coleoptera, Carabidae, Pterostichus).pp. 133-151 in J.L. Gressitt (Ed.),Pacific Basin biogeography. Bishop Museum Press, Honolulu (Symp. 10th Pacific Sci. Congr. 1961). 563 pp. (95 refs., 9 figs., 2 tabs.)

38. Ball, G.E. 1966. A revision of the North American species of the subgenus Cryobius Chaudoir (Pterostichus, Carabidae, Coleoptera). Opusc. ent. Suppl. 28. 166 pp. (101 refs., 210 figs., 63 tabs., keys)

39. Ball, G.E. and T.L. Erwin. 1969. A taxonomic synopsis of the tribe Loricerini (Coleoptera: Carabidae). Can. J. Zool. 47(5): 877-907. (39 refs., 62 figs., 4 tabs., key)

40. Banfield, A.W.F. 1954. Preliminary investigation of the barren ground Caribou. Part II. Life history, ecology and utilization. Can. Wildl. Serv. Wildl. Man. Bull. Ser. 1, 10B. 112 pp. (0 ref., 31 figs., 22 tabs.) [External parasites, pp. 33-37]

41. Bang-Haas, A. 1896. Lepidoptera Groenlandica. Vidensk. Meddr dansk. naturh. Foren. 48: 178-195. (In Danish) (text refs.)

42. Banks, N. 1919a. Part B: Neuropteroid insects. The Neuropteroid Insects of the Canadian Arctic Expedition 1913-1918. Rep. Can. Arctic Expedition, 1913-1918. Vol. III: Insects. 5 pp. (0 ref., 4 figs.)

43. Banks, N. 1919b. Part H: Spiders, Mites and Myriapods. The Acarina

collected by the Canadian Arctic Expedition, 1913-1918. Rep. Can. Arctic Expedition 1913-1918, Vol. III: Insects. pp. 11-13. (0 ref., 2 figs.)

44. Banks, N. 1923a. Trichoptera. p. 146 in W.L. McAtee (Ed.), A biological survey of the Pribilof Islands, Alaska. Part II, Insects, arachnids and chilopods of the Pribilof Islands, Alaska. N. Am. Fauna 46: 129-255. (text refs.)

45. Banks, N. 1923b. Mecoptera. p. 158 in W.L. McAtee (Ed.), A biological survey of the Pribilof Islands, Alaska. Part II, Insects, arachnids and chilopods of the Pribilof Islands, Alaska. N. Am. Fauna 46: 129-255. (0 ref., 1 fig.)

46. Banks, N. 1923c. Arachnida. pp. 237-239 in W.L. McAtee (Ed.), A biological survey of the Pribilof Islands, Alaska. Part II, Insects, arachnids and chilopods of the Pribilof Islands, Alaska. N. Am. Fauna 46: 129-255. (text refs., 7 figs.)

47. Barber, H.G. 1947. Revision of the genus Nysius in the United States and Canada (Hemiptera Heteroptera: Lygaeidae). J. Wash. Acad. Sci. 37(10): 354-366. (text refs., 1 fig., key)

48. Barca, E. 1937. Mikrolepidopteren aus Nordostgrönland. (Zool. Results Norwegian Sci. Exp. to East Greenland. VI). Skr. Svalbard Ishavet 71: 31-37, pl. III. (24 refs., 1 pl.)

49. Barlow, C.A. 1955. The fecundity of Aedes hexodontus Dyar (Culicidae) in the laboratory. Can. J. Zool. 33(5): 420-427. (13 refs., 5 figs., 1 tab.)

50. Barnes, W. and F.H. Benjamin. 1924. The North American and Greenland species of Sympistis (Lepidoptera). Can. Ent. 56(9): 212-215. (text refs., key)

51. Barnes, W. and A.W. Lindsey. 1921. The Pterophoridae of America, north of Mexico. Contr. nat. Hist. Lepid. N. Am. 4: 281-478, pls.XLI-LIV,Index. (text refs., 14 pls., keys)

52. Barr, A.R. and P.R. Ehrlich. 1957. A survey of the mosquitoes of the vicinity of Kotzebue, Alaska. Ladd Air Force Base, Alaska. U. S. Air Force. Arctic Aeromed. Lab. Tech. Note AAL-TN-57-9. (0 ref.)

53. Barr,A.R. and P.R. Ehrlich. 1958. Mosquito records from the Chukchi Sea coast of northwestern Alaska. Mosquito News 18(1): 12-14. (0 ref.)

54. Basden, E.B. 1956. Drosophilidae (Diptera) within the Arctic Circle. I. General Survey. Trans. R. ent. Soc. Lond. 108(1): 1-20. (24 refs.)

55. Batzli, G.O. and J. Brown. 1976. RATE [Research on arctic tundra environments]. The influence of grazing on arctic tundra ecosystems. Arct. Bull. 2(9): 153-160. (0 ref., 6 figs.)

56. Baumann, R.W. and A.R. Gaufin. 1970. The Capnia projecta complex of western North America (Plecoptera: Capniidae). Trans. Am. ent. Soc. 96(3): 435-468. (27 refs., 57 figs., keys)

57. Beckel, W.E. 1954a. The lack of autolysis of the flight muscles of Aedes communis (De Geer) (Culicidae) in the laboratory. Mosquito News 14(3): 124-127. (4 refs., 3 figs.)

58. Beckel, W.E. 1954b. The identification of adult female Aedes mosquitoes (Diptera, Culicidae) of the black-legged group taken in the field at Churchill, Manitoba. Can. J. Zool. 32 (4): 324-330. (3 refs., 3 figs., key)

59. Beckel, W.E. 1954c. Studies of the biology of the Aedes of northern Canada (Culicidae): I. Preliminary investigation of development in the egg. Def. Res. Bd Can. Northern Lab. Tech. Pap. 6 7 pp. (7 refs.)

60. Beckel, W.E. 1954d. Studies of the biology of the Aedes of northern Canada (Culicidae): II. Preliminary observations on a hatching stimulus for the eggs. Def. Res. Bd Can. Northern Lab. Tech. Pap. 7. 2 pp. (5 refs., 1 tab.)

61. Beckel, W.E. 1954e. Studies of the biology of the Aedes of northern Canada (Culicidae): V. Laboratory rearing of the adults. Def. Res. Bd Can. Northern Lab. Tech. Pap. 10. 7 pp. (1 ref., 3 tabs.)

62. Beckel, W.E. 1954f. Studies of the biology of the Aedes of northern Canada (Culicidae): VI. Collection of northern Aedes mosquito eggs from field-collected females. Def. Res. Bd Can. Northern Lab. Tech. Pap. 11. 4 pp. (0 ref., 1 fig.)

63. Beckel, W.E. 1955a. Oviposition site preference of Aedes mosquitoes (Culicidae) in the laboratory. Mosquito News 15(4): 224-228. (6 refs., 2 tabs.)

64. Beckel, W.E. 1955b. The presence of chitin in the transparent cuticle of the mosquito egg (Aedes hexodontus Dyar). Def. Res. Bd Can. Northern Lab. Tech. Mem. 5. 3 pp. (2 refs.)

65. Beckel, W.E. 1955c. A method of separating fertilized from non-fertilized mosquito eggs. Def. Res. Bd Can. Northern Lab. Tech. Mem. 6. 2 pp. (0 ref.)

66. Beckel, W.E. 1958a. Investigations of permeability, diapause, and hatching in the eggs of the mosquito Aedes hexodontus Dyar. Can. J. Zool. 36(4): 541-554. (29 refs., 1 fig.,1 tab.)

67. Beckel, W.E. 1958b. Observations on the rearing of larvae, pupae, and adults of some Aedes mosquitoes of northern Canada. Can. J. Zool. 36(5): 797-808. (6 refs., 4 tabs.)

68. Beckel, W.E. and C.A. Barlow. 1954. Studies of the biology of the Aedes of northern Canada (Culicidae): III. Field oviposition of Aedes communis (de Geer) with a method of separating the eggs from the substrate. Def. Res. Bd Can. Northern Lab. Tech. Pap. 8. 18 pp. (13 refs., 5 figs., 1 appendix)

69. Beckel, W.E. and T.P. Copps. 1954a. Studies of the biology of the Aedes of northern Canada (Culicidae): IV. Laboratory rearing of northern Aedes mosquito larvae and pupae. Def. Res. Bd Can. Northern Lab. Tech. Pap. 9. 11 pp. (2 refs., 6 tabs.)

70. Beckel, W.E. and T.P. Copps. 1954b. Notes on the rearing of Chaoborus sp. in the Churchill, Manitoba region. Def. Res. Bd Can. Northern Lab. Tech. Pap. 12. 4 pp. (0 ref., 1 fig.)

71. Beckel, W.E. and T.P. Copps. 1955a. A study of the diapause in the eggs of northern Aedes (Culicidae). Def. Res. Bd Can. Northern Lab. Rep. 5. 19 pp. (4 refs., 15 tabs.)

72. Beckel, W.E. and T.P. Copps. 1955b. An analysis of factors of importance to the rearing of northern mosquito larvae. Def. Res. Bd Can. Northern Lab. Rep. 6. 21 pp. (3 refs., 12 tabs.)

73. Beckel, W.E. and T.P. Copps. 1955c. Laboratory rearing of the adults of northern Aedes mosquitoes (Culicidae). Def. Res. Bd Can. Northern Lab. Rep. 7. 9 pp. (2 refs., 1 fig., 2 tabs.)

74. Beckel, W.E. and F.P. Donovan. 1955. Experimental fly bars for use in the sub-arctic transition forests and the arctic barrens. Def. Res. Bd Can. Northern Lab. Tech. Mem. 4. 6 pp. (0 refs., 4 figs.)

75. Becker, T. 1907. Beschreibung von 3 neuen Dipteren aus Ost-Grønland Meddr Grønland 29: 411-414. [Appendix to Nielsen 1907]. (0 ref.)

76. Behan, V.M. 1978. Diversity, distribution and feeding habits of North American arctic soil Acari. Ph.D. thesis, Dept. Ent., Macdonald College of McGill University. 428 pp.

77. Behan, V.M. and S.B. Hill. 1978. Feeding habits and spore dispersal of oribatid mites in the North American arctic. Rev. Ecol. Biol. Sol. 15(4): 497-516. (59 refs., 4 tabs., 2 appendixes)

78. Beirne, B.P. 1954a. New records and new synonymies for North American leafhoppers (Homoptera: Cicadellidae). Can. Ent. 86 (11): 495-498. (21 refs., 5 figs.)

79. Beirne, B.P. 1954b. A new genus and new species of deltocephaline leafhoppers from Canada and Alaska (Homoptera: Cicadellidae). Can. Ent. 86(12): 548-553. (3 refs., 5 figs., key)

80. Beirne, B.P. 1955. New species of leafhoppers from Canada (Homoptera: Cicadellidae). Can. Ent. 87(9): 373-375. (0 ref.)

81. Beirne, B.P. and D.A. Young, jr. 1953. The North American species of Cicadula (Homoptera: Cicadellidae). Can. Ent. 85(6): 215-226. (20 refs., 68 figs., keys)

82. Belicek, J. 1976a. Coccinellidae of western Canada and Alaska... M.Sc. thesis, Univ. Alta. 408 pp. [see Belicek 1976b]

83. Belicek, J. 1976b. Coccinellidae of western Canada and Alaska with analyses of the transmontane zoogeographic relationships between the fauna of British Columbia and Alberta (Insecta: Coleoptera: Coccinellidae). Quaest. ent. 12(4): 283-409. (269 refs., 1 + 16 + 193 figs., 2 tabs., keys)

84. Bell, R. 1884. Observations on the geology, minerology, zoology and botany of the Labrador coast, Hudson's Strait and Bay. Geol. Surv. Can. A. Rep. 1882-1884. [see Lyman 1884; Smith 1884]

85. Bell, R. 1885. Appendix II, Partial list of insects. pp. 26DD-27DD in Observations on the geology, zoology and botany of Hudson's Strait and Bay, made in 1885. Geol. Surv. Can. A. Rep. Pt. DD. 27 pp. (0 ref.)

86. Bell, R. 1901. Appendix 3. List of Lepidoptera taken in Baffin Land by Dr. Robert Bell in 1897. p. 38M in Report of an exploration on the northern side of Hudson Strait. Geol. Surv. Can. A. Rep. n.s. 11 (M). 38 pp. (0 ref.)

87. Benjamin, F.H. 1933. Notes on Phalaenidae from arctic America, supplemented with material from high altitudes. *Pan-Pacif. Ent.* 9(2): 53-62. (text refs.)

88. Bennett, G.F. and C.D. MacInnes. 1972. Blood parasites of geese of the McConnell River, N.W.T. *Can. J. Zool.* 50(1): 1-4. (22 refs., 1 tab.)

89. Bennett, G.F. and C.W. Sabrosky. 1962. The nearctic species of the genus *Cephenemyia* (Diptera, Oestridae). *Can. J. Zool.* 40(3): 431-448. (32 refs., 7 figs., 1 tab., key)

90. Benson, R.B. 1958. Arctic sawflies and the open habitat. *Proc. 10th int. Congr. Ent.* (Montreal 1956) 1: 693. (4 refs.)

91. Benson, R.B. 1959. Further studies on the Fenusini (Hymenoptera: Tenthredinidae). *Proc. R. ent. Soc. Lond.* 28(5-6): 90-92. (2 refs., 4 figs.)

92. Benson, R.B. 1962. Holarctic sawflies (Hymenoptera: Symphyta). *Bull. Br. Mus. nat. Hist.* (Ent.) 12(8): 381-409. (57 refs.)

93. Bequaert, J.C. 1941. Additions and corrections to the revision of North American Vespinae (Entomologica americana, 1932). Second paper. *Bull. Brooklyn ent. Soc.* 36(3): 111-117. (42 refs.)

94. Bergman, R.D. and D.V. Derksen. 1977. Observations on arctic and red-throated loons at Storkersen Point, Alaska. *Arctic* 3(1): 41-51. (14 refs., 4 tabs.)

95. Bertram, G.C.L. 1935. The low temperature limit of activity of Arctic insects. *J. anim. Ecol.* 4(1): 35-42. (10 refs., 1 fig., 1 tab., 1 appendix)

96. Bethune, C.J.S. 1870-1881. Insects of the northern parts of British America. From Kirby's Fauna Boreali-Americana: Insecta. 1870. *Can. Ent.* 2: 76-82; 89-93; 105-110; 142-145; 168-176(Coleoptera). 1871. *Can. Ent.* 3: 27-32; 88-94; 114-116; 134-137; 155-156; 172-176; 186-192; 211-217; 227-233 (Coleoptera). 1872. *Can. Ent.* 4: 31-36; 52-57; 93-96; 111-118; 151-155; 175-179; 196-198; 201-205 (Coleoptera). 1873. *Can. Ent.* 5: 96-99; 115-117; 129-132; 193-196; 210-213 (Coleoptera). 1875. *Can. Ent.* 7: 109-113 (Coleoptera); 129-132 (Orthoptera, "Neuroptera" [Odonata, Plecoptera], Trichoptera); 156-159 (Coleoptera additions; Hymenoptera). 1877. *Can. Ent.* 9: 148-156 (Hymenoptera). 1878. *Can. Ent.* 10: 116-118 (Hymenoptera); 137-139 (Hemiptera); 213-217 (Hemiptera, Homoptera, Lepidoptera). 1879. *Can. Ent.* 11: 146-154 (Lepidoptera). 1881. *Can. Ent.* 13: 162-170 (Lepidoptera, Diptera, "Homaloptera" [Dipt.], Aphaniptera [Siphonaptera]). (text refs.)

97. Bickley, W.A. 1976. Notes on the distribution of Alaskan mosquitoes. *Mosquito Systematics* 8(3): 232-236.(19 refs.)

98. Bickley, W.E. and E.G. MacLeod. 1956. A synopsis of the nearctic Chrysopidae with a key to the genera (Neuroptera). *Proc. ent. Soc. Wash.* 58(4): 177-202. (41 refs., 6 figs., keys)

99. Bierle, D.A. 1971. Dynamics of freshwater macrobenthic communities in Arctic Alaska ponds and lakes. pp. 228-233 *in* J. Brown and S. Bowen (Eds.), The Structure and Function of the Tundra Ecosystem, Vol. I. U.S. Tundra Biome Program, U.S. International Biological Program. (0 ref., 3 figs.)

100. Bierle, D.A. 1972. Production and energetics of chironomid larvae in ponds of the arctic coastal tundra. pp. 182-186 in Proc. 1972 Tundra Biome Symp. (15 refs., 4 figs., 2 tabs.)

101. Bissett, J. 1977. Preliminary observations on the vascular plants, fungi and insects of Polar Bear Pass, Bathurst Is., N.W.T. Unpub. report submitted to National Museum of Natural Sciences, Ottawa. 39 pp. (12 appendixes)

102. Blair, K.G. 1930. Coleoptera from Greenland. Oxford University Greenland Expedition, 1928. Ann. Mag. nat. Hist. 10(5): 394-400. (7 refs., 1 tab.)

103. Blair, K.G. 1933. Coleoptera collected by the Oxford University Expedition to Akpatok Island, Ungava Bay, Aug.-Sept. 1931. Ann. Mag. nat. Hist. 10(12): 93-96. (0 ref.)

104. Blake, W., Jr. 1974. Studies of glacial history in arctic Canada. II. Interglacial peat deposits on Bathurst Island. Can. J. Earth Sci. 11(8): 1025-1042. (93 refs., 6 figs., 3 tabs.)

105. Blake, W., Jr., and J.V. Matthews, Jr. 1979. New data on an interglacial peat deposit near Makinson Inlet, Ellesmere Island, District of Franklin. Geol. Surv. Can. Pap. 79-1A: 157-164. (18 refs., 6 figs., 2 tabs.)

106. Blanton, F.S., B.V. Travis, N. Smith, and C.N. Husman. 1950. Control of adult mosquitoes in Alaska with aerial sprays. J. econ. Ent. 43(3): 347-350. (3 refs., 2 tabs.)

107. Bliss, L.C. 1972a. IBP high arctic ecosystem study, Devon Island. Arctic 25(2): 158-161. (0 ref.)

108. Bliss, L.C. 1972b. Devon Island research 1971. pp. 269-275 in F.E. Wielgolaski and T. Rosswall (Eds.), Biological productivity of tundra, Tundra Biome. Oslo-Stockholm, IBP Tundra Biome Steering Committee. 319 pp. (0 ref., 2 figs.)

109. Bliss, L.C. 1972c. Devon Island I.B.P. Project, high arctic ecosystem, project report 1970 and 1971. Dept. Botany, Univ. Alta. 413 pp. [see Ryan 1972]

110. Bliss, L.C. 1973. Devon Island, Canada, high arctic ecosystem. Biol. Conservation 3: 229-231. (0 ref., 1 fig.)

111. Bliss, L.C. 1975a. Truelove Lowland - A high arctic ecosystem. pp. 51-85 in T.W.M. Cameron and L.W. Billingsley (Eds.), Energy flow - its biological dimensions. A summary of the IBP in Canada: 1964-1974. Royal Society of Canada, Ottawa. 319 pp. (59 refs., 2 figs., 7 tabs.)

112. Bliss, L.C. 1975b. Devon Island, Canada. pp. 17-60 in T. Rosswall and O.W. Heal (Eds.), Structure and function of tundra ecosystems. Ecol. Bull. 20 (Stockholm). 450 pp. (50 refs., 8 figs., 13 tabs.)

113. Bliss, L.C. (Ed.) 1977. Truelove lowland, Devon Island, Canada: A high arctic ecosystem. Univ. Alta Press, Edmonton. 714 pp. [see Addison, Minns, Pattie, Procter, Ryan, Ryan and Hergert, Whitfield]

114. Bliss, L.C., G.M. Courtin, D.L. Pattie, R.R. Riewe, and D.W.A. Whitfield. 1973. Arctic tundra ecosystems. A. Rev. Ecol.

Syst. 4: 359-399. (274 refs.)

115. Böcher, J. 1971. Preliminary studies on the biology and ecology of Chlamydatus pullus (Reuter) (Heteroptera: Miridae) in Greenland. Meddr Grønland 191(3). 29 pp. (37 refs., 11 figs., 6 pls., 3 tabs.)

116. Böcher, J. 1972a. Feeding biology of Nysius groenlandicus (Zett.) (Heteroptera: Lygaeidae) in Greenland. Meddr Grønland 191(4). 41 pp. (31 refs., 12 figs., 8 tabs., 5 pls.)

117. Böcher, J. 1972b. [Arctic insects - insects in Greenland I, II.] Grønland 1972 (5): 133-143; 1972 (6): 169-182. (In Danish) (27 refs., 14 figs.)

118. Böcher, J. 1975. [Dispersal of Nysius groenlandicus (Heteroptera, Lygaeidae) in Greenland.] Ent. Meddr 43(2): 105-109. (In Danish, English summary) (2 refs., 2 figs.)

119. Böcher, J. 1976. Population studies on Nysius groenlandicus (Zett.) (Heteroptera: Lygaeidae) in Greenland with particular reference to climatic factors, especially the snow cover. Vidensk. Meddr dansk naturh. Foren. 139: 61-89. (38 refs., 20 figs., 5 tabs.)

120. Böcher, J. 1978. Biology and ecology of the arctic-alpine bug Nysius groenlandicus (Zett.) (Het., Lygaeidae) in Greenland. Norw. J. Ent. 25(1): 72. (4 refs.)

121. Bödvarsson, H. 1960. One new genus and three new species of Collembola from Alaska. Opusc. ent. 25(1-2): 43-51. (6 refs., 3 figs.)

122. Bohnsack, K.K. 1968. Distribution and abundance of tundra arthropods in the vicinity of Pt. Barrow, Alaska. Final Rep., Arct. Inst. N. Am., Project ONR-308 and ONR-321. 111 pp. Mimeo.

123. Bohnsack, K.K. 1973. Distribution of Oribatids near Barrow, Alaska. pp. 71-74 in M. Daniel and B. Rosicky (Eds.), Proc. 3rd int. Congr. Acarol. Akademia Publishing House, Prague. 837 pp. (6 refs., 1 tab.)

124. Borkent, A. 1979. Systematics and bionomics of the species of the subgenus Schadonophasma Dyar and Shannon (Chaoborus, Chaoboridae, Diptera). Quaest. ent. 15(2): 122-255. (229 refs., 40 figs., 59 tabs.)

125. Bradley, J.C. 1956. The distribution of northeastern insects. Ent. News 67(10): 257-261. (0 ref.)

126. Braendegaard, J. 1934. Preliminary statement of the synonymy of certain Greenland spiders. Vidensk. Meddr dansk naturh. Foren. 98: 47-48. (0 ref.)

127. Braendegaard, J. 1935. Insects and Arachnids III. Araneae. The Scoresby Sound Committee's 2nd East Greenland Expedition in 1932 to King Christian IX's land. Meddr Grønland 104(16): 16-18. (4 refs.)

128. Braendegaard, J. 1936. Revisal of spiders from Ellesmereland collected by the Second Arctic Expedition of the "Fram". Norsk ent. Tidsskr. 4(3): 128-130. (8 refs.)

129. Braendegaard, J. 1937. Spiders (Araneina) from southeast Greenland. Meddr Grønland 108(4). 14 pp. (12 refs., 6 figs.)

130. Braendegaard, J. 1938. Aeronautic spiders in the Arctic. Meddr Grønland 119(5). 9 pp. (20 refs., 3 figs.)

131. Braendegaard, J. 1939. I. Supplementary list to "Spiders (Araneina) from southeast Greenland." II. Arctosa alpigena Dol. and Arctosa insignita Th. (Fam. Lycosidae). Meddr Grønland 108(7). 11 pp. (13 refs., 5 figs., 2 tabs.)

132. Braendegaard, J. 1940. I. Spiders (Araneina) from northeast Greenland between lats. 70°25' and 76°50'N. II. On the possibility of a reliable determination of species of the females of the genus Erigone. Meddr Grønland 125(8). 29 pp. (32 refs., 26 figs., 1 tab.)

133. Braendegaard, J. 1946. The zoology of East Greenland. Araneina. Meddr Grønland 121(15). 128 pp. (123 refs., 49 figs., 14 tabs.)

134. Braendegaard, J. 1960. The spiders (Araneida) of Peary land, north Greenland. Meddr Grønland 159(6). 24 pp. (31 refs., 5 figs., 5 tabs.)

135. Braendegaard, J., K.L. Henriksen, and R. Spärck. 1935. Insects and arachnids. The Scoresby Sound Committee's 2nd East Greenland Expedition in 1932 to King Christian IX's Land. Meddr Grønland 104(16). [See Braendegaard, Henriksen, Spärck]

136. Breddin, G. 1902. Die Hemipteren und Siphunculaten des arktischen Gebietes. Fauna arct. 2: 529-560. (84 + text refs.)

137. Brinck, P. 1940. Beitrag zur Kenntnis der Wasserkafer Grönlands. Opusc. ent. 5(2-4): 37-41. (8 refs., 8 figs.)

138. Brooks, A.R. 1944. A review of the North American species of Linnaemya sens. lat. (Diptera, Tachinidae). Can. Ent. 76(10): 193-206. (text refs., 11 figs., keys)

139. Brooks, A.R. and L.A. Kelton. 1967. Aquatic and semiaquatic Heteroptera of Alberta, Saskatchewan and Manitoba (Hemiptera). Mem. ent. Soc. Can. 51. 92 pp. (105 refs., 138 figs., keys)

140. Brooks, G.S. 1942. A revised checklist of the butterflies of Manitoba. Can. Ent. 74(2): 31-36. (0 ref.)

141. Brooks, S.T., W.J. Holland, A. Avinoff, and C. Heinrich. 1935. The exploration of Southampton Island, Hudson Bay, by George Miksch Sutton. Part II. Zoology; section 5. Insects (Lepidoptera) and invertebrates. Mem. Carnegie Mus. 12, Part II, sect. 5. 35 pp. [see Heinrich, Holland and Avinoff]

142. Brown, J. 1971. The U.S. Tundra Biome Program. Proc. 22nd Alaska Sci. Conf.:28. [abstract only]

143. Brown, J. 1972. Summary of the 1971 U.S. Tundra Biome Program. pp. 306-313 in F.E. Wielgolaski and T. Rosswall (Eds.), Tundra Biome. Proc. IV Int. Meeting on the Biological Productivity of Tundra, Leningrad. (0 ref., 2 figs.)

144. Brown, J. (Ed.). 1975. Ecological investigations of the tundra biome in the Prudhoe Bay region, Alaska. Univ. Alaska, Biol. pap. spec. rep. 2. 215 pp. [see MacLean; Norton et al.]

145. Brown, J. and S. Bowen (Eds.). 1971. The structure and function of the tundra ecosystem, Vol. 1. U.S. Tundra Biome Program, U.S.

International Biological Program. 282 pp. [See Bierle, Edwards, MacLean 1971a, Philip 1971a, Schmoller]

146. Brown, J. and H.N. Coulombe. 1969. Analysis of the structure and function of the wet tundra ecosystem at Barrow, Alaska. U.S. IBP Tundra Biome Barrow Workshop, Univ. Colorado.

147. Brown, J. and G.C. West. 1970. Tundra biome research in Alaska. The structure and function of cold-dominated ecosystems. U.S. IBP-Tundra Biome Report 70-1. 297 pp. [Now published by MacLean, Norton and others (q.v.)]

148. Brown, J.M. 1932. Report on the Collembola collected by the Oxford University Expedition to Hudson Strait, 1931. Ann. Mag. nat. Hist. 10(10): 330-340. (20 refs., 7 figs., 1 tab.)

149. Brown, J.M. 1937. On some Collembola from Iceland and Greenland. Ann. Mag. nat. Hist. 10(2): 514-520. (19 refs.)

150. Brown, W.J. 1931. New species of Coleoptera II. Can. Ent. 63(5): 115-122. (text refs.)

151. Brown, W.J. 1932. New species of Coleoptera, III. Can. Ent. 64(1): 3-12. (0 ref., key)

152. Brown, W.J. 1937. The Coleoptera of Canada's eastern Arctic. Can. Ent. 69(5): 106-111. (text refs.)

153. Brown, W.J. 1938. Some new Canadian Chrysomelidae. Can. Ent. 70 (2): 35-38. (text refs.)

154. Brown, W.J. 1940a. Some new species of Cantharidae and Chrysomelidae (Coleoptera). Can. Ent. 72(8): 161-166. (text refs., 3 figs.)

155. Brown, W.J. 1940b. Some new and poorly known species of Coleoptera. Can. Ent. 72(9): 182-187. (text refs., key)

156. Brown, W.J. 1942a. The American species of Phytodecta Kby (Coleoptera, Chrysomelidae). Can. Ent. 74(6): 99-105. (text refs., 4 figs., key)

157. Brown, W.J. 1942b. The American species of Entomoscelis and Hippuriphila (Coleoptera, Chrysomelidae). Can. Ent. 74(9): 172-176. (7 + text refs., 7 figs., key)

158. Brown, W.J. 1949. On the American species of Lyperopherus Mots. (Coleoptera: Carabidae). Can. Ent. 81(9): 231-232. (text refs., key)

159. Brown, W.J. 1952. Some species of Phytophaga (Coleoptera). Can. Ent. 84(11): 335-342. (15 refs.)

160. Brown, W.J. 1956. The new world species of Chrysomela L.(Coleoptera: Chrysomelidae). Can. Ent. Suppl. 3. 54 pp. (62 refs., 40 figs., key)

161. Brown, W.J. 1961. Notes on North American Chrysomelidae (Coleoptera). Can. Ent. 93(11): 967-977. (text refs., key)

162. Brown, W.J. 1962. The American species of Chrysolina Mots. (Coleoptera: Chrysomelidae). Can. Ent. 94(1): 58-74. (text refs., 14 figs., key)

163. Brown, W.J. 1965a. A hydrophilid new to North America (Coleoptera: Hydrophilidae). Coleopts Bull. 19(1): 31. (text refs.)

164. Brown, W.J. 1965b. Distribution of the beetles of the Eastern Canadian Arctic. pp. 29-32 in A.H. Macpherson (Ed.), Summary of papers and discussions at meeting of Arctic Circle, Ottawa, 1964. Arct. Circ. 16(2): 14-34. ([2] refs.)

165. Brown, W.J. 1966. Chrysomelinae and Curculionidae (Coleoptera): descriptions and notes. Can. Ent. 98(8): 855-859. (10 refs., key)

166. Brown, W.L. 1955. The ant Leptothorax muscorum (Nylander) in North America. Ent. News 66(2): 43-50. (text + 7 refs.)

167. Brues, C.T. 1903. A monograph of the North American Phoridae. Trans. Am. ent. Soc. 29(4): 331-404. Pl. V-IX. (25 refs., 60 figs., keys)

168. Brues, C.T. 1919. Part G: Hymenoptera and Plant Galls. The parasitic Hymenoptera collected by the Canadian Arctic Expedition 1913-1918. Rep. Can. Arctic Expedition 1913-1918. Vol. III: Insects. pp. 21-24. (0 ref.)

169. Bruggemann, P.F. 1958. Insects and environments of the high arctic. Proc. 10th int. Congr. ent. (Montreal 1956) 1: 659-702. (8 refs., 1 tab.)

170. Brundin, L. 1940. Zwei neue nordische Atheta-Arten aus der islandica-Gruppe (Col. Staphylinidae). Vorlaufige Mitteilung. Ent. Tidskr. 61: 131-133. (0 ref.)

171. Brust, R.A. 1966. Intersexes in Aedes nigripes (Zett.) (Diptera: Culicidae). Mosquito News 26(4): 512-514. (7 refs., 4 figs.)

172. Brust, R.A. 1968. Temperature-induced intersexes in Aedes mosquitoes: comparative study of species from Manitoba. Can. Ent. 100(8): 879-891. (25 refs., 38 figs., 1 tab.)

173. Brust, R.A. 1971. Laboratory mating in Aedes diantaeus and Aedes communis (Diptera: Culicidae). Ann. ent. Soc. Am. 64(1): 234-237. (19 refs., 1 fig., 1 tab.)

174. Bruun, H. and M. von Schantz. 1950. [A contribution to the knowledge of Brenthis improba Btl. ssp. improbula Bryk(Lepid.).] Notul.ent. 29(3-4): 83-89. (In Finnish,English summary)(8 refs., 4 figs.)

175. Bryant, O. 1935. Notes on the diurnal Lepidoptera of the Canadian Arctic... Bull. Brooklyn ent. Soc. 30: 1-10, 42-62, 115-118. [See Leussler and Bryant 1935]

176. Buckle, D.J. and J.H. Redner. 1964. The nearctic species of the Xysticus labradorensis subgroup (Araneae: Thomisidae). Can. Ent. 96(8): 1138-1142. (7 refs., 13 figs., key)

177. Bunnell, F.L., S.F. MacLean, Jr., and J. Brown. 1975. Barrow, Alaska, U.S.A. pp. 73-124 in T. Rosswall and O.W. Heal (Eds.), Structure and function of tundra ecosystems. Ecol. Bull. (Stockholm) 20. 450 pp. (79 refs., 14 figs., 7 tabs.)

178. Burgoyne, W.E. 1978. Mosquito controls during construction of the Trans-Alaska pipeline. Mosquito News 38(1): 25-35. (2 refs., 4 figs., 4 tabs.)

179. Burton, A.N. and J. McLintock. 1970. Further evidence of western encephalitis infection in Saskatchewan mammals and birds and in reindeer in northern Canada. Can. Vet. J. 99(11): 232-235. (20 refs., 1 tab.)

180. Bushnell, J.H. 1970. Preliminary biological reconnaissance of aquatic habitats, Cumberland Peninsula, Baffin Island. Am. Zool. 10(4): 554. [abstract only]

181. Bushnell, J.H. and E.R. Byron. 1979. Morphological variability and distribution of aquatic invertebrates (principally Crustacea) from the Cumberland Peninsula and Frobisher Bay Regions, Baffin Island, N.W.T., Canada. Arct. Alp. Res.11(2): 159-177. (47 refs., 6 figs., 4 tabs.)

182. Butler, A.G. 1877. Description of a new species of Argynnis from arctic America. Entomologist's mon. Mag. 13: 206. (0 refs.)

183. Callot, J. 1954. Moustiques au Groenland. pp. 25-26 in Expéditions polaires françaises. Expédition arctique 1948-1953. Rapports préliminaires, 25. (4 refs.)

184. Campbell, J.M. 1973. A revision of the genus Tachinus (Coleoptera: Staphylinidae) of North and Central America. Mem. ent. Soc. Can. 90. 137 pp. (84 refs., 26 maps, 189 figs., key)

185. Campbell, J.M. 1978. A revision of the North American Omaliinae (Coleoptera: Staphylinidae). 2. The tribe Coryphiini. Mem. ent. Soc. Can. 106: 21-87. (63 refs., 69 figs., 10 maps)

186. Campbell, J.M. 1979. A revision of the genus Tachyporus Gravenhorst (Coleoptera: Staphylinidae) of North and Central America. Mem. ent. Soc. Can. 109. 95 pp. (58 refs., 106 figs., 23 maps, 1 colour pl., key)

187. Canadian Arctic Expedition 1913-1918, Report. Vol. III: Insects. 1912-1922. Parts A-L. King's Printer, Ottawa. [Indexed here by author. Introduction (Hewitt); A. Collembola (Folsom); B. Neuropteroid insects (Banks); C. Diptera (Alexander, Dyar, Malloch); D. Mallophaga and Anoplura (Baker; Ferris and Nuttall); E. Coleoptera, (Swaine, Fall, Leng, Sherman); F. Hemiptera (Van Duzee); G. Hymenoptera and Plant Galls (MacGillivray, Brues, Sladen, Felt); H. Spiders, mites and myriapods (Emerton, Banks, Chamberlin); I. Lepidoptera (Gibson); J. Orthoptera (Walker); K. Insect life... (Johansen); L. General Index] [The report, Vol. III, includes also a preface, contents, errata]

188. Carpenter, G.D.H. 1938. Notes on insects collected in West Greenland by the Oxford University Greenland Expedition, 1936. With descriptions of a new species of Angitia (Hymenoptera, Ichneumonidae) by A. Roman, and of Fannia (Diptera, Anthomyidae) by J. E. Collin. Ann. Mag. nat. Hist. (11), 1(5): 529-553. (0 ref.)

189. Carpenter, G.D.H. and J.E. Collin. 1937. Notes on insects collected in Northwest Greenland by the Oxford University Ellesmere Land Expedition, 1934-1935, by G.D.H. Carpenter, with description of a new species of Rhamphomyia (Dipt.) by J.E. Collin. Ann. Mag. nat. Hist. 10(20): 401-409. (text refs., 1 fig., 1 tab.)

190. Carpenter, G.D.H. and A. Holm. 1939. Insecta and Araneae collected in the Kangerdlugsuak region of East Greenland by the British Expedition, 1935-36. Ann. Mag. nat. Hist. 11(3): 60-80. (22

refs., 3 figs.)

191. Carpenter, S.J. and W.J. LaCasse. 1955. Mosquitoes of North America north of Mexico. Univ. Calif. Press, Berkeley and Los Angeles. 360 pp. (771 refs., 288 figs., 127 pls., keys)

192. Carr, F.S. 1930. New Canadian Coleoptera, I. Can. Ent. 62(12): 278-279. (0 ref., 3 figs., key)

193. Carr, F.S. 1932. New arctic Coleoptera. Can. Ent. 64(8): 191-192. (0 ref.)

194. Cary, M. 1907. On the diurnal Lepidoptera of the Athabaska and Mackenzie Region, British America. Proc. U.S. natn. Mus. 31: 425-457. (19 refs.)

195. Cassagnau, P. 1972. Un Collembola adapté à la prédation: Cephalotoma grandiceps (Reuter). Nouv. Rev. Ent. 2(1): 5-12. (11 refs., 4 figs.)

196. Caudell, A.N. 1915. Podisma frigida Boh. in Alaska. Can. Ent. 47 (5): 160. (0 ref.)

197. Challet, G.L. and K.K. Bohnsack. 1968. Distribution and abundance of Collembola at Pt. Barrow, Alaska. Pedobiologia 8(2): 214-222. (6 refs., 6 tabs.)

198. Chamberlin, R.V. 1919. Part H: Spiders, Mites and Myriapods. The Chilopoda collected by the Canadian Arctic Expedition, 1913-1918. Rep. Can. Arctic Expedition 1913-1918. Vol. III: Insects. pp. 15-22. (0 ref., 5 figs.)

199. Chamberlin, R.V. 1923. Chilopoda. pp. 240-244 in W.L. McAtee (Ed.), A biological survey of the Pribilof Islands, Alaska. Part II, Insects, arachnids and chilopods of the Pribilof Islands, Alaska. N. Am. Fauna 46: 129-255. (text refs.)

200. Chamberlin, R.V. 1946. On the chilopods of Alaska. Ann. ent. Soc. Am. 39(2): 177-189. (7 refs., 6 figs., key)

201. Chamberlin, R.V. 1949. On some American spiders of the family Erigonidae. Ann. ent. Soc. Am. 41(4): 483-562. (text refs., 162 figs., text tabs.)

202. Chamberlin, R.V. and W.J. Gertsch. 1958. The spider family Dictynidae in America north of Mexico. Bull. Am. Mus. nat. Hist. 116(1): 1-152. (text refs., 37 figs., 47 pls., keys)

203. Chamberlin, R.V. and W. Ivie. 1947. The spiders of Alaska. Bull. Univ. Utah 37(10). 103 pp. (26 refs., 99 figs.)

204. Chant, D.A. 1963. The subfamily Blattisocinae Garman (=Aceosejinae Evans) (Acarina: Blattisocidae Garman) (=Aceosejidae Baker and Wharton) in North America, with descriptions of new species. Can. J. Zool. 41(2): 243-305. (65 refs., 149 figs., keys)

205. Chant, D.A. and R.I.C. Hansell. 1971. The genus Amblyseius (Acarina, Phytoseiidae) in Canada and Alaska. Can. J. Zool. 49(5): 703-758. (33 refs., 242 figs., keys)

206. Chapman, H.C., J.R. Gorham, and T. Fukuda. 1973. Thelohania (Nosematidae: Microsporida) in Aedes mosquitoes of Alaska. Mosquito News 33(3): 465-466. (5 refs., 1 tab.)

207. Charlton, M.N. 1971. Aspects of the energy budget of Pseudodiamesa arctica kept in the laboratory. M.Sc. thesis, Univ. Toronto. 99 pp.

208. Chernov, Yu. I. 1968. [The distribution and abundance of Collembola in the tundra zone.] Ent. Obozr. 47(1): 80-85. (In Russian) [Translation in Ent. Rev. 47(1): 47-49.] (19 refs.)

209. Children, J.G. 1836. Articulata. Appendix 3, pp. 532-542 in Back, Sir G., Narrative of the Arctic Land Expedition to the mouth of the Great Fish River, and along the shores of the arctic ocean in the years 1833, 1834 and 1835. J. Murray, London. 663 pp.

210. Chillcott, J.G. 1960. A revision of the Nearctic species of Fanniinae (Diptera: Muscidae). Can. Ent. Suppl. 14. 295 pp. (338 refs., 9 + 289 figs. + 61 maps, 1 tab., keys)

211. Chillcott, J.G. 1962. A revision of the Platypalpus juvenis complex in North America (Diptera: Empididae). Can. Ent. 94(2): 113-143. (7 refs., 50 figs., 5 maps, key)

212. China, W.E. 1934. Hemiptera collected by the Oxford University Expedition to West Greenland, 1928. Ann. Mag. nat. Hist. 13(10): 330-333. (text refs., 2 pls.)

213. Christiansen, K.A. 1951. Notes on Alaskan Collembola. II. Three new species of arctic Collembola. Psyche, Camb. 58(4): 125-140. (7 refs., 12 pls.)

214. Christiansen, K. 1958. The nearctic members of the genus Entomobrya (Collembola). Bull. Mus. comp. Zool. Harv. 118(7): 440-545, pls. 1-24. (text + 54 refs., 24 pls., keys)

215. Circumpolar Conference on Northern Ecology, Ottawa. 1975. Proceedings. Natn. Res. Council Canada. [see Hobbie 1975]

216. Clark, A.H. 1935. Arctic butterflies. Smithsonian Rep. 1934: 267-296, 7 pls. [Publ. 3305] (0 ref., 7 pls.(=56 figs.))

217. Clausen, P.J. and E.F. Cook. 1971. A revision of the nearctic species of the tribe Parydrini (Diptera: Ephydridae). Mem. Am. ent. Soc. 27. 150 pp. (85 refs., 234 figs., keys)

218. Clay, T. 1949. Species of the genus Saemundssonia (Mallophaga) from the Sterninae. Am. Mus. Novit. 1409. 25 pp. (2 refs., 31 figs., 7 tabs., key)

219. Clement, L.E. 1975. The ecology of Tipula carinifrons Holm. (Diptera: Tipulidae) in the arctic coastal tundra of northern Alaska. Unpub. thesis, Univ. Alaska, Fairbanks. 96 pp.

220. Cloudsley-Thompson, J.L. 1949. On spiders from north-east Greenland, 1948. Ann. Mag. nat. Hist. 12(2): 398-400. (4 refs.)

221. Cole, F.R. 1921. Diptera from the Pribilof Islands, Alaska. pp. 169-177 in G.D. Hanna et al., Insects of the Pribilof Islands, Alaska. Proc. Calif. Acad. Sci. ser 4, 11(14): 153-195. (text refs., 4 figs.)

222. Collin, J.E. 1930. A revision of the Greenland species of the anthomyid genus Limnophora sens. lat. (Diptera), with figures of the male genitalia of these and many other palaearctic

species. Trans. ent. Soc. Lond. 78: 255-281, pl. XV-XXVII (text refs., 13 pls., key)

223. Collin, J.E. 1931. The Oxford University expedition to Greenland, 1928. Diptera (Orthorrhapha Brachycera and Cyclorrhapha) from Greenland. Ann. Mag. nat. Hist. 10(7): 67-91. (text refs., 27 figs.)

224. Collin, J.E. 1935. Diptera (Cyclorrhapha) from Akpatok Island, Ungava Bay, Canada. Results of the Oxford University Hudson Strait Expedition, 1931. Ann. Mag. nat. Hist. 10(15): 369-381. (text refs., 11 figs.)

225. Collin, J.A. 1937. On a collection of Diptera from East Greenland. Ann. Mag. nat. Hist. 10(19): 594-595. (text refs., 1 tab.)

226. Collin, J.E. 1951. A new arctic species of Fucellia (Diptera, Anthomyidae) with maculated wings. Ent. Meddr 26(2): 187-190. (0 ref., 2 figs.)

227. Cook, E.F. 1955. A contribution toward a monograph of the family Scatopsidae (Diptera). Part I. A revision of the genus Rhegmoclema Enderlein (=Aldrovandiella Enderlein) with particular reference to the North American species. Ann. ent. Soc. Am. 48(4): 240-251. (13 refs., 5 pls., keys)

228. Cook, E.F. 1956a. A contribution toward a monograph of the Scatopsidae (Diptera). Part IV, the genus Swammerdamella (Enderlein). Ann. ent. Soc. Am. 49(1): 15-29. (text + 8 refs., 6 figs., keys)

229. Cook, E.F. 1956b. A contribution toward a monograph of the Scatopsidae (Diptera). Part VI, the genera Scatopse Geoffroy and Holoplagia Enderlein. Ann. ent. Soc. Am. 49(6): 593-611. (text + 13 refs., 7 figs., key)

230. Cook, E.F. 1956c. The nearctic Chaoborinae (Diptera: Culicidae). Univ. Minn. Agr. exp. Stn Tech. Bull. 218. 102 pp. (64 refs., 34 figs., keys)

231. Coolidge, K.R. 1908. The North American species of the genus Erebia (Lepidoptera). Ent. News 19(2): 73-75. (0 ref.)

232. Coquillett, D.W. 1902. New Diptera from North America. Proc. U.S. natn. Mus. 25(1280): 83-126. (0 ref.)

233. Corbet, P.S. 1964. Autogeny and oviposition in arctic mosquitoes. Nature, Lond. 203(4945): 669. (11 refs., 1 tab.)

234. Corbet, P.S. 1965a. Reproduction in mosquitoes of the high arctic. Proc. 12th int. Congr. Ent. (London 1964): 817-818. (3 refs., 2 figs.)

235. Corbet, P.S. 1965b. An insect emergence trap for quantitative studies in shallow ponds. Can. Ent. 97(8): 845-848. (4 refs., 1 fig.)

236. Corbet, P.S. 1966a. Diel patterns of mosquito activity in a high arctic locality: Hazen Camp, Ellesmere Island, N.W.T. Can. Ent. 98(12): 1238-1252. (18 refs., 5 figs.)

237. Corbet, P.S. 1966b. Parthenogenesis in caddisflies (Trichoptera). Can. J. Zool. 44(5): 981-982. (13 refs.)

238. Corbet, P.S. 1967. Facultative autogeny in arctic mosquitoes.

Nature, Lond. 215(5105): 662-663. (16 refs., 1 tab.)

239. Corbet, P.S. 1972. The microclimate of arctic plants and animals on land and in fresh water. Acta arct. 18. 43 pp.(128 refs., 23 figs.)

240. Corbet, P.S. and H.V. Danks. 1973. Seasonal emergence and activity of mosquitoes in a high arctic locality. Can. Ent. 105(6): 837-872. (66 refs., 21 figs., 7 tabs.)

241. Corbet, P.S. and H.V. Danks. 1975. Egg-laying habits of mosquitoes in the high arctic. Mosquito News 35(1): 8-14. (18 refs., 4 figs.)

242. Corbet, P.S. and A.E.R. Downe. 1966. Natural hosts of mosquitoes in northern Ellesmere Island. Arctic 19(2): 153-161. (21 refs., 1 fig., 1 tab.)

243. Corpuz-Raros, L.A. and E.F. Cook. 1974. A revision of North American Capitophorus Van der Goot and Pleotrichophorus Börner (Homoptera: Aphididae). Smithson. Contribs Zool. 156. 143 pp. (191 refs., 494 figs., 1 tab., keys, 3 appendixes)

244. Cotton, M.J. 1979a. A collection of spiders from north-east Greenland. Arctic 32(1): 71-75. (5 refs., 3 tabs.)

245. Cotton, M.J. 1979b. Records of insects from north-east Greenland. Entomologist's mon. Mag. 113: 213-217. (1 ref., 1 tab.)

246. Covell, C.V., Jr. 1970. A revision of the North American species of Scopula (Lepidoptera, Geometridae). Trans. Am. ent. Soc. 96: 101-221. (116 refs., 99 figs., key)

247. Craig, P.C. and J. Wells. 1976. Life history notes for a population of slimy sculpin (Cottus cognatus) in an Alaskan arctic stream. J. Fish. Res. Bd Can. 33(7): 1639-1642. (12 refs., 2 figs., 2 tabs.)

248. Cranston, P.S. 1979. The development stages of Limnophyes globifer (Lundström) (Diptera: Chironomidae). Hydrobiologia 67(1): 19-25. (13 refs., 7 figs., text tabs.)

249. Cresson, E.T. 1877. Notes on the species belonging to the subfamily Ichneumonides, found in America north of Mexico. Trans. Am. ent. Soc. 6: 129-212. (text refs., keys)

250. Cresson, E.T. 1918. New North American Diptera (Scathophagidae). Ent. News 29(4): 133-137. (text refs.)

251. Cresson, E.T. 1930. Studies in the dipterous family Ephydridae, paper III. Trans. Am. ent. Soc. 56(2): 93-131. (text refs.)

252. Crosby, C.R. 1934. I. Some Arachnida from Southampton Island. The exploration of Southampton Island, Hudson Bay, by G.M. Sutton. Mem. Carnegie Mus. 12, part II, sect. 4: 1-2. (text refs.)

253. Crosby, C.R., C.P. Alexander, J.R. Malloch, W.D. McIlroy, Jr., and H.S. Peters. 1934. The exploration of Southampton Island, Hudson Bay, by George Miksch Sutton. Part II, Zoology; section 4. Spiders and insects (in part). Mem. Carnegie Mus. 12, part II, sect. 4. 37 pp. [See Alexander, Crosby, Malloch, McIlroy, Peters]

254. Crosby, C.R. and S.C. Bishop. 1928. Revision of the spider genera Erigone, Eperigone and Catabrithorax (Erigoneae). Bull. N.Y. St. Mus. 278: 3-94. (text refs., 151 figs., keys)

255. Crosby, C.R. and S.C. Bishop. 1933. American spiders: Erigoneae, males with cephalic pits. Ann. ent. Soc. Am. 26(1): 105-182. (text refs., 236 figs., keys)

256. Crosson du Cormier, A. 1977. Les Boloria du domaine arctique. Nécessité d'une nouvelle division spécifique (Nymphalidae). Alexanor 10(1): 35-47. (text refs., 18 + 1 figs.)

257. Curran, C.H. and C.L. Fluke. 1926. Revision of the nearctic species of Helophilus and allied genera. Trans. Wis. Acad. Sci. Arts Lett. 22: 207-821. (text refs., 40 figs., keys)

258. Curtis, J. 1835. Descriptions & c. of the insects brought home by Commander James Clark Ross, R.N., F.R.S., & c. pp. lix-lxxx in J.C. Ross, Account of the objects in the several departments of Natural History, seen and discovered during the present expedition. pp. i-cxliv in J. Ross, Appendix to the narrative of a second voyage in search of a north-west passage, and of a residence in the arctic regions during the years 1829, 1830, 1831, 1832, 1833. Webster, London. [reprinted 1969, Greenwood, N.Y.] (text refs., 15 figs.)

259. Curtis, J. 1836. Insekten. pp. 286-294 in W.F. Erichson, Zoologische Resultate von J. Ross' zweiter Nordpol-Reise. Arch. Naturgesch. 2: 281-302. (0 ref.)

260. Custer, T.W. and F.A. Pitelka. 1977. Demographic features of a Lapland Longspur population near Barrow, Alaska. Auk 94(3): 505-525. (45 refs., 5 figs., 7 tabs.)

261. Custer, R.W. and F.A. Pitelka. 1978. Seasonal trends in summer diet of the Lapland Longspur near Barrow, Alaska. Condor 80(3): 295-301. (19 refs., 3 figs., 1 tab.)

262. Dahl, C. 1960. Trichocera arctica Lundström (Dipt. Nematoc.) new to North America. Opusc. ent. 25(1-2): 150-152. (4 refs., 2 figs.)

263. Dahl, C. 1967. Notes on the taxonomy and distribution of arctic and subarctic Trichoceridae (Dipt. Nem.) from Canada, Alaska and Greenland. Opusc. ent. 32(1-2): 49-78. (41 refs., 83 figs.)

264. Dahl, C. 1970. Distribution, phenology and adaptation to arctic environment in Trichoceridae (Diptera). Oikos 21(2): 185-202. (169 refs., 7 figs., 2 tabs.)

265. Danks, H.V. 1971a. A note on the early season food of arctic migrants. Can. Fld Nat. 85(1): 71-72. (6 refs., 1 tab.)

266. Danks, H.V. 1971b. Spring and early summer temperatures in a shallow arctic pond. Arctic 24(2): 113-123. (12 refs., 2 figs., 2 tabs.)

267. Danks, H.V. 1971c. Overwintering of some North Temperate and Arctic Chironomidae. I. The winter environment. Can. Ent. 103(4): 589-604. (50 refs., 9 figs., 1 tab.)

268. Danks, H.V. 1971d. Overwintering of some North Temperate and Arctic Chironomidae. II. Chironomid biology. Can. Ent. 103(12): 1875-1910. (105 refs., 9 figs., 13 tabs.)

269. Danks, H.V. 1978. Modes of seasonal adaptation in the insects. I. Winter Survival. Can. Ent. 110(11): 1167-1205. (351 refs., 6 tabs.)

270. Danks, H.V. (Ed.) 1979a. Canada and its insect fauna. Mem. ent. Soc. Can. 108. 573 pp. (ca 3230 refs., 77 figs., 59 tabs., 2 appendixes)

271. Danks, H.V. 1979b. Characteristic modes of adaptation in the Canadian insect fauna. pp. 548-566 in H.V. Danks (Ed.), Canada and its insect fauna. Mem. ent. Soc. Can. 108. 573 pp. (115 refs.)

272. Danks, H.V. and J.R. Byers. 1972. Insects and arachnids of Bathurst Island, Canadian arctic archipelago. Can. Ent. 104(1): 81-88. (24 refs., 3 figs., 1 tab.)

273. Danks, H.V. and P.S. Corbet. 1973a. A key to all stages of Aedes impiger and A. nigripes (Diptera: Culicidae), with a description of first instar larvae and pupae. Can. Ent. 105(3): 367-376. (24 refs., 6 figs., key)

274. Danks, H.V. and P.S. Corbet. 1973b. Sex ratios at emergence of two species of high arctic Aedes (Diptera: Culicidae). Can. Ent. 105(4): 647-651. (23 refs., 3 tabs.)

275. Danks, H.V. and D.R. Oliver. 1972a. Seasonal emergence of some high arctic Chironomidae (Diptera). Can. Ent. 104(5): 661-686. (61 refs., 16 figs., 2 tabs.)

276. Danks, H.V. and D.R. Oliver. 1972b. Diel periodicities of emergence of some high arctic Chironomidae (Diptera). Can. Ent. 104(6): 903-916. (47 refs., 7 figs.)

277. Darsie, R.F., Jr. 1957. Notes on American mosquito pupae II. The Aedes (Ochlerotatus) punctor subgroup, with a key to known nearctic Aedes pupae (Diptera, Culicidae). Ann. ent. Soc. Am. 50(6): 611-620. (13 refs., 6 figs., 1 tab., keys)

278. Dasch, C.E. 1964. Ichneumon-flies of America north of Mexico: 5. Subfamily Diplazontinae. Mem. Am. ent. Inst. 3. 304 pp. (text refs., 153 figs., 111 maps, keys)

279. Dasch, C.E. 1971. Ichneumon-flies of America north of Mexico: 6. Subfamily Mesochorinae. Mem. Am. ent. Inst. 16. 376 pp. (text refs., 462 figs., 131 maps, keys)

280. Davies, D.M. 1959. The parasitism of black flies (Diptera, Simuliidae) by larval water mites mainly of the genus Sperchon. Can. J. Zool. 37(3): 353-369. (28 refs., 5 figs., 7 tabs.)

281. Davies, D.M. and B.V. Peterson. 1956. Observations on the mating, feeding, ovarian development and oviposition of adult black flies (Simuliidae, Diptera). Can. J. Zool. 34(6): 615-655. (125 refs., 7 figs., 4 tabs.)

282. Davies, I.J. 1975. Selective feeding in some arctic Chironomidae. Verh. Int. Verein. theor. angew. Limnol. 19(4): 3149-3154. (5 refs., 2 figs., 1 tab.)

283. Davis, D.H.S. 1936. A reconnaissance of the fauna of Akpatok Island, Ungava Bay. J. anim. Ecol. 5(2): 319-332. (16 refs., 2 figs.)

284. Degerbøl, M. 1937. A contribution to the investigation of the fauna of the Blosseville Coast, East Greenland, with special reference to zoogeography. Meddr Grønland 104(19). 35 pp. (43 refs., 1 fig., 1 pl., 1 tab.)

285. Deichmann, H. 1896. [Insects of East Greenland. Brief observations

on insect life.] Meddr Grønland 19: 97-104. (In Danish)(0 ref.)

286. Deichmann, H. and W. Lundbeck. 1896. [Insects of East Greenland.] Meddr Grønland 19: 95-120. (In Danish) [see Deichmann 1896; Lundbeck 1896]

287. de Lesse, H. 1948. Zoologie, 4e note. Macrolepidoptera. Expéditions Polaires françaises, missions Paul-Emile Victor. Ann. Soc. ent. Fr. 118: 51-78. (43 refs., 38 figs.)

288. de Lesse, H. 1949. Observations sur l'époque d'apparition et le comportement des lépidoptères du Groenland occidental et remarques sur leur homochromie. Expéditions polaires françaises, mission P.E. Victor. Revue fr. Lepidopt. 12(9-10): 163-169.

289. de Lesse, H. 1950a. Notes zoologiques et botaniques sur l'ouest du Groenland. Expéditions polaires françaises, missions P.E.Victor Terre Vie 20(4): 175-201, pls. A-C. (29 refs., 3 pls., 1 tab.)

290. de Lesse, H. 1950b. Entomologie et botanique. pp. 76-80 in P.E. Victor et al. Rapport préliminaire de la campagne au Groenland 1949. Expéditions polaires françaises. Expédition arctique, 1949. Publications no. 10. (text ref.)

291. de Lesse, H. 1950c. Coléoptères du Groenland. Expédition polaire française. Zoologie (3e note). Entomologiste 6(3): 73-76. (0 ref.)

292. de Lesse, H. 1952. Zoologie, 6e note: Biogéographie des lépidoptères du Groenland. Expéditions polaires françaises, missions P.E. Victor. Ann. Soc. ent. Fr. 119: 97-116. (30 refs., 4 figs., 1 tab.)

293. de Lesse, H. and C. Granger. 1952. Hymenoptera. Expéditions polaires françaises (missions P.E. Victor). Zoologie (8e note). Bull. Soc. ent. Fr. 57(4): 53-58. (text refs.)

294. de Lesse, H., R. Strenzke, and A. Thienemann. 1951. Expéditions polaires françaises (missions P.E. Victor). Zoologie (5e note). Bull. Soc. ent. Fr. 56(4): 53-61. (8 refs., 1 fig.)

295. de Lesse, H. and P. Viette. 1949. Microlepidoptera. Expéditions polaires françaises (missions P.E. Victor). Campagne 1949 au Groenland. Zoologie - première note. Ann. Soc. ent. Fr. 115: 81-92. (6 refs., 14 figs.)

296. Dondale, C.D. and J.H. Redner. 1975. The fuscomarginatus and histrio groups of the spider genus Philodromus in North America (Araneida: Thomisidae). Can. Ent. 107(4): 369-384. (33 refs., 3 maps, 40 figs., key)

297. Dondale, C.D. and J.H. Redner. 1978. The insects and arachnids of Canada. Part 5. The crab spiders of Canada and Alaska (Araneae: Philodromidae and Thomisidae). Agric. Can. Publ. 1663. 255 pp. (121 refs., 66 maps, 725 figs., keys)

298. Dondale, C.D. and J.H. Redner. 1979. Revision of the wolf spider genus Alopecosa Simon in North America (Araneae: Lycosidae). Can. Ent. 111(9): 1033-1055. (63 refs., 56 figs., 5 maps, key)

299. Dondale, C.D., A.L. Turnbull, and J.H. Redner. 1964. Revision of the

nearctic species of Thanatus C.L. Koch (Araneae: Thomisidae). Can. Ent. 96(4): 636-656. (19 refs., 59 figs., 1 tab., keys).

300. Dos Passos, C.F. 1947. Erebia youngi Holland, its subspecies and distribution (Lepidoptera, Satyridae). Am. Mus. Novit. 1348: 1-4. (text refs.)

301. Dos Passos, C.F. 1949a. New butterflies from Mount McKinley National Park, Alaska, with a review of Erebia rossii (Rhopalocera, Satyridae). Am. Mus. Novit. 1389: 1-17. (9 + text refs., 28 figs.)

302. Dos Passos, C.F. 1949b. The distribution of Oeneis taygete Geyer in North America with descriptions of new subspecies (Lepidoptera, Satyridae). Am. Mus. Novit. 1399: 1-21. (text + 11 refs., 16 figs.)

303. Dos Passos, C.F. 1958. The satyrid butterflies of northwestern North America. Proc. 10th int. Congr. Ent. (Montreal 1956) 1: 673-681. (134 refs., 1 tab.)

304. Douce, G.K. 1973. The population dynamics and community analysis of the Acarina of the arctic coastal tundra. Unpub. M.Sc. thesis, Univ. Georgia, Athens. 69 pp.

305. Douce, G.K. 1976a. Biomass of soil mites (Acari) in arctic coastal tundra. Oikos 27(2): 324-330. (17 refs., 3 figs., 5 tabs.)

306. Douce, G.K. 1976b. The role of soil fauna in litter decomposition processes in arctic tundra, evaluated by field and systems analysis techniques. Unpub. Ph.D. thesis, Univ. Georgia, Athens. 116 pp.

307. Douce, G.K. and D.A. Crossley. 1977. Acarina abundance and community structure in an arctic coastal tundra. Pedobiologia 17(1): 32-42. (36 refs., 1 fig., 6 tabs.)

308. Douce, G.K. and D.P. Webb. 1978. Indirect effects of soil invertebrates on litter decomposition: elaboration via analysis of a tundra model. Ecol. Model. 4(4): 339-359. (53 refs., 4 tabs., 1 appendix)

309. Douglas, M.M. 1978. Behavioural and biophysical convergence in the thermo-regulatory strategies of high arctic butterflies. Trans. Kans. Acad. Sci. 81(2): 142-143. [abstract only]

310. Downes, J.A. 1958. The genus Culicoides (Diptera: Ceratopogonidae) in Canada; an introductory review. Proc. 10th int. Congr. Ent. (Montreal 1956) 3: 801-808. (43 refs., 3 figs., 1 appendix)

311. Downes, J.A. 1962. What is an arctic insect? Can. Ent. 94(2): 143-162. (77 refs., 1 tab.)

312. Downes, J.A. 1964. Arctic insects and their environment. Can. Ent. 96(1-2): 280-307. (71 refs., 2 pls., 2 tabs.)

313. Downes, J.A. 1965. Adaptations of insects in the arctic. A. Rev. Ent. 10: 257-274. (77 refs.)

314. Downes, J.A. 1966. The Lepidoptera of Greenland: Some geographic considerations. Can. Ent. 98(11): 1135-1144. (21 refs., 1 tab.)

315. Downes, J.A. 1970. The feeding and mating behaviour of the

specialised Empidinae (Diptera); observations on four species of Rhamphomyia in the high arctic and a general discussion. Can. Ent. 102(7): 769-791. (39 refs., 5 figs., 4 tabs.)

316. Drake, C.J. 1952. Alaskan Saldidae (Hemiptera). Proc. ent. Soc. Wash. 54(3): 145-148. (text refs.)

317. Drake, C.J. 1957. An undescribed shore-bug from Manitoba (Hemiptera Saldidae). Bull. sth Calif. Acad. Sci. 56(3): 142-143.

318. Dyar, H.G. 1899. Notes on Alaskan Arctiidae. Ent. News 10(5): 130-131. (0 ref.)

319. Dyar, H.G. 1919. Part C: Diptera. The mosquitoes collected by the Canadian Arctic Expedition 1913-1918. Rep. Can. Arctic Expedition 1913-1918. Vol. III. pp. 31-33. (0 ref., 2 figs.)

320. Dyar, H.G. and R.C. Shannon. 1925. The mosquitoes of Peary's North Pole Expedition of 1908 (Diptera, Culicidae). J. Wash. Acad. Sci. 15(4): 77-78. (text refs.)

321. Ebner, J.A. and C.D. Ferris. 1977. A new subspecies of Colias palaeno (Linnaeus) from Baffin Island, N.W.T., Canada (Pieridae). J. Res. Lepid. 16(3): 155-161. (7 refs., 10 figs.)

322. Edwards, F.W. 1931. Oxford University Greenland Expedition, 1928. - Diptera Nematocera. Ann. Mag. nat. Hist. 10(8): 617-618. (text refs.)

323. Edwards, F.W. 1933. Oxford University Expedition to Hudson's Strait, 1931: Diptera Nematocera. With notes on some other species of the genus Diamesa. Ann. Mag. nat. Hist. 10(12): 611-620. (0 ref., 2 figs.)

324. Edwards, F.W. 1935. Diptera Nematocera from East Greenland. Ann. Mag. nat. Hist. 10(15): 467-473. (text refs., 2 figs.)

325. Edwards, J.S. 1971. Arthropod fallout on Alaskan snowfields. Proc. 22nd Alaska Sci. Conf.: 30. [abstract only]

326. Edwards, J.S. 1971. Abundance of soil arthropods. pp. 130-133 in J. Brown and S. Bowen (Eds.), The structure and function of the tundra ecosystem, Vol. I. U.S. Tundra Biome Program, U.S. International Biological Program. (0 ref., 2 tabs.)

327. Edwards, J.S. 1972. Soil invertebrates in North American arctic and alpine tundra. Proc. 1972 Tundra Biome Symp. (Univ. Wash. 1972 U.S. Tundra Biome. Ecosyst. Anal. Stud. U.S. IBP U.S. Arctic Res. Prog.) pp. 139-143. (9 refs., 5 tabs.)

328. Edwards, J.S. 1972. Arthropod fallout on Alaskan snow. Arct. Alp. Res. 4(2): 167-176. (5 refs., 4 tabs.)

329. Edwards, J.S. and P.C. Banko. 1976. Arthropod fallout and nutrient transport: A quantitative study of Alaskan snowpatches. Arct. Alp. Res. 8(3): 237-245. (22 refs., 1 fig., 5 tabs.)

330. Edwards, W.H. 1877. Catalogue of the diurnal Lepidoptera of America, north of Mexico. Trans. Am. ent. Soc. 6: 1-68. (text refs.)

331. Edwards, W.H. 1879. Insects, diurnal Lepidoptera. pp. 155-157 in L. Kumlien et al.,Contributions to the natural history of arctic America. Bull. U.S. natn. Mus. 15. 179 pp. (0 ref.)

332. Edwards, W.H. 1883. Descriptions of new species of diurnal Lepidoptera, found in British America and the United States. Can. Ent. 15(1): 32-36. (text refs.)

333. Edwards, W.H. and E.W. Nelson. 1887. Part IV. Insects: Introduction. Diurnal Lepidoptera collected in Alaska. pp. 323-330 in E.W. Nelson et al., Report upon the natural history collections made in Alaska between ...1877 and 1881. Washington, D.C. (text refs.)

334. Egede, H. 1741. [New survey of old Greenland, or natural history and description of the situation, climate and characteristics of Old Greenland.] J.C. Groth, Copenhagen. (In Danish) [also published in English, 1745, A description of Greenland.]

335. Ehrlich, P.R. 1952. The distribution and subspeciation of Erebia rossii Curtis (Lepidoptera: Satyridae). Trans. Am. ent. Soc. 78(2): 75-88. (10 refs.)

336. Ehrlich, P.R. 1958a. Lepidoptera collected in the tundra-taiga ecotone at Kotzebue, Alaska. Ent. News 69(1): 17-19. (4 refs.)

337. Ehrlich, P.R. 1958b. Problems of arctic-alpine insect distribution as illustrated by the butterfly genus Erebia (Satyridae). Proc. 10th int. Congr. Ent. (Montreal 1956) 1: 683-686. (2 refs., 8 figs.)

338. Eichlin, T.D. and H.B. Cunningham. 1978. The Plusiinae (Lepidoptera: Noctuidae) of America north of Mexico, emphasizing genitalic and larval morphology. U.S. Dep. Agric. Tech. Bull. 1567. 122 pp. (119 refs., 19 maps, 227 figs., keys)

339. Ellis, R.A. and R.A. Brust. 1973. Sibling species delimitation in the Aedes communis (Degeer) aggregate (Diptera: Culicidae). Can. J. Zool. 51(9): 915-959. (367 refs., 8 figs., 16 tabs., keys)

340. El-Moursy, A.A. 1970. The taxonomy of the nearctic species of the genus Byrrhus Linnaeus (Coleoptera: Byrrhidae). Quaest.ent. 6 (4): 327-338. (9 refs., 24 figs., key)

341. Elwes, H.J. 1889. A revision of the genus Argynnis. Trans. ent. Soc. Lond. 1889: 535-575. (text refs.)

342. Elwes, H.J. 1903. On a collection of Lepidoptera from arctic America. Trans ent. Soc. Lond. 1903(3): 239-243. (text refs., 1 pl.)

343. Elwes, H.J. 1904. Lepidoptera from arctic America. Appendix III, pp. 271-275 in D.T. Hanbury, Sport and travel in the northland of Canada. MacMillan Co., N.Y. 319 pp. [Same as Elwes 1903]. (text refs., 12 figs.)

344. Emerson, K.C. 1950. The genus Lagopoecus (Philopteridae: Mallophaga) in North America. J. Kans. ent. Soc. 23(3): 97-101. (text refs., 2 pls.)

345. Emerton, J.H. 1918. Collecting notes for northern spiders. Ottawa Naturalist (later Can. Fld Nat.) 32(1): 13. (0 ref.)

346. Emerton, J.H. 1919a. Part H: Spiders, Mites and Myriapods. The spiders collected by the Canadian Arctic Expedition, 1913-1918. Rep. Can. Arctic Expedition 1913-1918. Vol. III: Insects. pp. 3-9. (0 ref., 20 figs.)

347. Emerton, J.H. 1919b. Notes on the spiders collected by Frits Johansen of the "Danmark Expedition" 1906-1908 to north-eastern Greenland, latitude 76 1/2° to 77° North. Vidensk. Meddr dansk naturh. Foren. 70: 143-145. (0 ref., 7 figs.)

348. Emerton, J.H. 1920. Catalogue of the spiders of Canada known to the year 1919. Trans. R. Can. Inst. 12: 309-338. (text refs.)

349. Emerton, J.H. 1921. Notes on Canadian and arctic spiders. Psyche, Camb. 28(5-6): 165-168. (text refs.)

350. Emerton, J.H. 1927. Recent finds of arctic spiders. Can. Fld Nat. 41(4): 88. (0 ref.)

351. Enderlein, G. 1935. Neue Simuliiden, besonders aus Afrika. Sitzungsberichte Ges. naturf. Freunde 1934: 358-364. (0 ref.)

352. Enderlein, G. 1936. Simuliologica I. Sitzungsberichte Ges. naturf. Freunde 1936: 113-130. (text refs.)

353. Erichson, W.F. 1836. Zoologische Resultate von J. Ross' zweiter Nordpol-Reise. Arch. Naturgesch. 2: 281-302. [see Curtis 1836]

354. Etzel, A.V. 1859. Uebersicht der Land-, Susswasser- und Ufer-Arthropoden Gronlands. Berl. ent. Z. 3: 134-157. [see Schiödte 1859]

355. Evans, G.O. 1955. A collection of mesostigmatid mites from Alaska. Bull. Br. Mus. nat. Hist. (Zool.) 2(9): 287-307. (18 refs., 42 figs., key)

356. Ewing, H.E. 1923. The dermanyssid mites of North America. Proc. U. S. natn. Mus. 62: 13(2459): 1-26. 1 pl. (text refs., 20 figs., keys)

357. Fabricius, O. 1780. Favna groenlandica, systematice sistens animalia Groenlandiae occidentalis hactenvs indigata, quoad nomen specificum, triviale, vernaculumque. Hafniae et Lipsiae, Gottlob Rothe. xvi + 452 pp. (text refs., 12 figs.)

358. Fain, A. and K. Hyland. 1974. The Listrophoroid mites in North America II. The family Listrophoridae Megnin and Trouessart (Acarina: Sarcoptiformes). Bull. Inst. r. Sci. nat. Belg. 31 (8). 69 pp. (32 refs., 65 figs., 1 tab., key)

359. Fall, H.C. 1900. List of a small collection of Coleoptera from arctic Alaska. Ent. News 11: 459-460. (0 ref.)

360. Fall, H.C. 1919. Carabidae and Silphidae. pp. 14-16 in Part E. Coleoptera. The Coleoptera collected by the Canadian Arctic Expedition 1913-1918. Rep. Can. Arctic Expedition 1913-1918. Vol. III: Insects. 27 pp. (0 ref.)

361. Fall, H.C. 1922. A review of the North American species of Agabus. John D. Sherman, Jr., Mt. Vernon, N.Y. 36 pp. (text refs., key)

362. Fall, H.C. 1923. A revision of the North American species of Hydroporus and Agaporus. Privately printed. 129 pp. (refs. in species list; keys)

363. Fauvel, A. 1878. Les Staphylinides de l'Amérique du Nord. Bull. Soc. Linn. Normandie 3(2): 167-266. (text refs., keys)

364. Felt, E.P. 1919. Part G: Hymenoptera and Plant Galls. The plant galls collected by the Canadian Arctic Expedition 1913-1918. Rep. Can. Arctic Expedition 1913-1918. Vol. III: Insects. pp. 37-38. (0 ref.)

365. Ferguson, D.C. 1978. Noctuoidea, Lymantriidae. In R.B. Dominick et al., The Moths of America North of Mexico, Fasc. 22.2. E.W. Classey Ltd. and The Wedge Entomological Research Foundation, London, England. 110 pp. + pl. (118 refs., 1 pl., 8 colour pls., 23 figs., keys)

366. Fernald, C.H. 1894. North Greenland microlepidoptera. Ent. News 5(5): 129-132. (0 ref.)

367. Ferns, P.N. 1977b. General ornithological notes. pp. 152-165 in G. H. Green and J.J.D. Greenwood (Eds.), Report of the Joint Biological Expedition to N.E. Greenland 1974. Dundee: Dundee University N.E. Greenland Expedition. 258 pp. (refs., 8 figs., 1 tab.)

368. Ferris, C.D. 1972. Polymorphism in two species of Alaskan Boloria (Nymphalidae). J. Res. Lepid. 11(4): 255-259. (1 ref., 2 figs.)

369. Ferris, C.D. 1974a. Distribution of arctic-alpine Lycaena phlaeas L. (Lycaenidae) in North America with designation of a new subspecies. Bull. Allyn Mus. 18: 1-13. (21 refs., 21 figs., 1 tab.)

370. Ferris, C.D. 1974b. Notes on arctic and subarctic collecting. J. Res. Lepid. 13(4): 249-264. (4 refs., 3 figs., text tab.)

371. Ferris, G.F. 1923. Mallophaga. p. 141 in W.L. McAtee (Ed.), a biological survey of the Pribilof Islands, Alaska. Part II, Insects, arachnids and chilopods of the Pribilof Islands, Alaska. N. Am. Fauna 46: 129-255. (text refs.)

372. Ferris, G.F. 1951. The sucking lice. Mem. Pacif. Coast. ent. Soc. 1. 320 pp. (text refs., 124 figs., keys)

373. Ferris, G.F. and G.H.F. Nuttall. 1919. Part D: Mallophaga and Anoplura. Anoplura of the Canadian Arctic Expedition, 1913-1918. Rep. Can. Arctic Expedition 1913-1918. Vol. III. Insects. p. 11. (0 ref.)

374. Fjellberg, A. 1978. New species of the genus Isotoma Bourlet, 1839 from North America (Collembola: Isotomidae). Ent. Scand. 9(2): 93-110. (2 refs., 86 figs., key)

375. Fluke, C.L. 1933. Revision of the Syrphus flies of America north of Mexico I. Trans. Wis. Acad. Sci. Arts Lett. 28: 63-128. (text refs., 44 figs., keys)

376. Fluke, C.L. 1949. Some Alaskan Syrphid flies, with descriptions of new species. Proc. U.S. natn. Mus. 100(3256): 39-54. (text refs., 11 figs., 1 tab., keys)

377. Fluke, C.L. 1952. The Metasyrphus species of North America (Diptera, Syrphidae). Am. Mus. Novit. 1590. 27 pp. (text refs., 54 figs., key)

378. Fluke, C.L. 1954. Two new North American species of Syrphidae, with notes on Syrphus (Diptera). Am. Mus. novit. 1690. 10 pp. (text refs., 20 figs., key)

379. Fluke, C.L. and F.M. Hull. 1946. Syrphid flies of the genus Cheilosia, subgenus Chilomyia in North America (Part II). Trans. Wis. Acad. Sci. Arts Lett. 36 (1944): 327-347. (text refs., 17 figs., keys)

380. Folsom, J.W. 1901. The distribution of holarctic Collembola. Psyche, Camb. 9: 159-162. (0 ref.)

381. Folsom, J.W. 1916. North American collembolous insects of the subfamilies Achorutinae, Neanurinae, and Podurinae. Proc. U.S. natn. Mus. 50(2134): 477-525, pls. 7-25. (117 refs., 250 figs., keys)

382. Folsom, J.W. 1917. North American collembolous insects of the subfamily Onychiurinae. Proc. U.S. natn. Mus. 53 (2222): 637-659, pls. 68-79. (96 refs., 101 figs., keys)

383. Folsom, J.W. 1919a. Part A: Collembola. Collembola of the Canadian Arctic Expedition, 1913-1918. Rep. Can. Arctic Expedition 1913-1918. Vol. III: Insects. 29 pp. (115 refs., 72 figs.)

384. Folsom, J.W. 1919b. Collembola from the Crocker Land Expedition 1913-1917. Bull. Am. Mus. nat. Hist. 41: 271-302, pls. XLI-XXIII. (90 refs., 60 figs.)

385. Folsom, J.W. 1937. Nearctic collembola or springtails, of the family Isotomidae. Bull. U.S. natn. Mus. 168. 144 pp. (190 refs., 461 figs., 3 tabs., keys)

386. Foote, B.A. 1971. Biology of Hedria mixta (Diptera: Sciomyzidae). Ann. ent. Soc. Am. 64(4): 931-941. (33 refs., 25 figs.)

387. Forbes, W.T.M. 1923. Lepidoptera. pp. 147-149 in W.L. McAtee (Ed.), A biological survey of the Pribilof Islands, Alaska. Part II, Insects, arachnids and chilopods of the Pribilof Islands, Alaska. N. Am. Fauna 46: 129-255. (text refs.)

388. Forbes, W.T.M. 1954. Lepidoptera of New York and neighboring states. Part III. Noctuidae. Mem. Cornell Univ. Agric. Exp. Stn 329. 433 pp. (text refs., 290 figs., keys)

389. Forbes, W.T.M. 1960. Lepidoptera of New York and neighboring states. Part IV. Agaristidae through Nymphalidae including butterflies. Mem. Cornell Univ. agric. Exp. Stn 371. 188 pp. (0 ref., 188 figs., keys)

390. Forsslund, K.H. 1932. Zur Kenntnis der Trichopteren Grönlands. Ent. Tidskr. 53(1): 56-59. (text refs., 1 fig.)

391. Forsslund, K.H. 1957. Notizen über Oribatei (Acari), I. Ark. Zool. n.s. 10(18): 583-593. (22 refs., 14 figs.)

392. Forsslund, K.H. 1964. Liochthonius muscorum n. sp. und L. lapponicus (Träg.) (Acari, Oribatei). Ent. Tidskr. 85 (3-4): 236-239. (9 refs., 5 figs.)

393. Fox, W.J. 1892. Report on the Hymenoptera collected in West Greenland. Proc. Acad. nat. Sci. Philad.: 133-135. (0 ref.)

394. Francoeur, A. 1974. Nouvelles données et remarques sur la répartition nordique de quelques formicides (Hyménoptères) néarctiques. Naturaliste Can. 101: 935-936. (8 refs.)

395. Franklin, H.J. 1913. The Bombidae of the New World I. Trans. Am. ent. Soc. 38: 177-486. (Plates follow Vol. 39, p. 200). (text refs., pls., text tabs., keys)

396. Freeman, R.B. 1956. Some insects from North Greenland. Entomologist's mon. Mag. 92(201): 330. (0 ref.)

397. Freeman, T.N. 1939. Notes on the distribution of *Plebeius aquilo* Bdv. with the description of a new race from Manitoba (Lepid. Lycaenidae). Can. Ent. 71(8): 178-180. (0 ref., 1 fig.)

398. Freeman, T.N. 1940. Two apparently new Canadian species of *Sparganothis*. Can. Ent. 72(10): 206-208. (0 ref., 2 figs.)

399. Freeman, T.N. 1948a. The arctic Lepidoptera of Baker Lake, Northwest Territories. Lepid. News 2(6): 63-65. (0 ref., 1 fig.)

400. Freeman, T.N. 1948b. The field season summary of North American Lepidoptera for 1948. 8. Far North. Lepid. News II, Suppl. xi-xii. (0 ref.)

401. Freeman, T.N. 1949a. The Northern Insect Survey in Canada and some environmental observations. 80th A. Rep. ent. Soc. Ont.: 1-3. (0 ref.)

402. Freeman, T.N. 1949b. The field season summary of North American Lepidoptera for 1949. 8. Far North. Lepid. News 3: 101-102. (0 ref.)

403. Freeman, T.N. 1950a. The field season summary of North American Lepidoptera for 1950. 8. Far North. Lepid.News 4: 106-107. (0 ref.)

404. Freeman, T.N. 1950b. The Northern Insect Survey for 1950. Arct. Circ. 3: 55-56. (0 ref.)

405. Freeman, T.N. 1951a. Northern Canada and some northern butterflies. Lepid. News. 5: 41-42. (0 ref.)

406. Freeman, T.N. 1951b. The Northern Insect Survey for 1951. Arct. Circ. 4(6): 85-87. (0 ref.)

407. Freeman, T.N. 1951c. The field season summary of North American Lepidoptera for 1951. 8. Far North. Lepid. News 5: 110. (0 ref.)

408. Freeman, T.N. 1952a. Some problems of insect biology in the Canadian arctic. Arctic 5(3): 175-177. (0 ref., 1 fig.)

409. Freeman, T.N. 1952b. Interim report of the distribution of the mosquitoes obtained in the Northern Insect Survey. Def. Res. Bd Can. Environ. Protect. Tech. Rep. 1. 2 pp. + 43 pp. maps. (0 ref., 43 figs., 1 tab.)

410. Freeman, T.N. 1953a. Interim report of the distribution of the tabanids obtained in the Northern Insect Survey. Def. Res. Bd Can. Environ. Protect. Tech. Rep. 4. 3 pp. + 35 maps. (0 ref., 35 figs.)

411. Freeman, T.N. 1953b. The Northern Insect Survey for 1952. Arct. Circ. 6(3): 30-32. (0 ref.)

412. Freeman, T.N. 1953c. The field season summary of North American

Lepidoptera for 1952. 8. Far North. Lepid. News 7: 118. (0 ref.)

413. Freeman, T.N. 1957. The northern insect survey, 1953-56. Arct. Circ. 10(1): 10-13. (0 ref.)

414. Freeman, T.N. 1958a. A historical account of insect collecting in northern Canada. Proc. 10th int. Congr. Ent. (Montreal 1956) 1: 613-617. (12 refs., 1 fig.)

415. Freeman, T.N. 1958b. The distribution of arctic and subarctic butterflies. Proc. 10th int. Congr. Ent. (Montreal 1956) 1: 659-672. (0 ref., 27 figs.)

416. Freeman, T.N. 1959. The Canadian Northern Insect Survey, 1947-1957. Polar Rec. 9(61): 299-307. (48 refs., 1 fig.)

417. Freeman, T.N. 1964a. Postglacial invasion routes of Canadian arctic butterflies. Can. Ent. 96(1-2): 116. (0 ref.)

418. Freeman, T.N. 1964b. ...And butterflies and beetles too! pp. 34-38 in I.N. Smith (Ed.), The unbelievable land. Queen's Printer, Ottawa. 140 pp. (0 ref.)

419. Freeman, T.N. 1972. A correlation of some butterfly distributions with geological formations. Can. Ent. 104(3): 443-444. (0 ref., 1 fig.)

420. Freeman, T.N. and W. Hovanitz. 1947. Field summary of Lepidoptera - 1947 season. 8. Far North. Lepid. News 1: 96. (0 ref.)

421. Freeman, T.N. and C.R. Twinn. 1955. Present trends and future needs of entomological research in northern Canada. Arctic 7(3-4): 275-283. [also pp. 163-171 in Arctic Research. Spec. Pub. Arct. Inst. N. Am. 2. (1955)] (32 + 24 refs., 1 fig.)

422. Frey, R. 1922. Vorarbeiten zu einer Monographie der Gattung Rhamphomyia Meig. (Dipt. Empididae). Notul. ent. 2(1): 1-10; (2): 33-45; (3): 65-77. (text refs., keys)

423. Friese, H. 1902. Die arktischen Hymenopteren, mit Ausschluss der Tenthrediniden. Fauna arct. 2: 439-488. (36 refs., 1 fig., 1 pl., 3 tabs.)

424. Friese, H. 1935. Apiden aus nordost-Grönland. (Zoological results of the Norwegian Scientific Expeditions to East Greenland. IV). Skr. Svalbard Ishavet 65: 3-10. (30 refs., 1 tab.)

425. Frison, T.H. 1919. Report on the Bremidae collected by the Crocker Land Expedition, 1913-1917. Bull. Am. Mus. nat. Hist. 41: 451-459. pl. xxiv. (10 refs., 2 + 2 figs.)

426. Frison, T.H. 1921. Hymenopterous insects of the family Bremidae from the Pribilof Islands, Alaska. pp. 185-187 in G.D. Hanna et al., Insects of the Pribilof Islands. Proc. Calif. Acad. Sci. ser. 4, 11(14): 153-195. (0 ref.)

427. Frison, T.H. 1923. Systematic and biological notes on bumblebees (Bremidae; Hymenoptera). Trans. Am. ent. Soc. 48(4): 307-326. (text refs.)

428. Frison, T.H. 1934. Four new species of stoneflies from North America: Plecoptera. Can. Ent. 66(2): 25-30. (0 ref., 16 figs.)

429. Frohne, W.C. 1953. Mosquito distribution in Alaska with especial reference to a new type of life cycle. Proc. 4th Alaska Sci. Conf.:171-177. (6 refs.)

430. Frohne, W.C. 1954. Mosquito distribution in Alaska with especial reference to a new type of life cycle. Mosquito News 14(1): 10-13. (6 refs.)

431. Frohne, W.C. 1955a. Characteristic saddle spines of northern mosquito larvae. Trans. Am. microsc. Soc. 74(3): 295-302. (6 refs., 24 figs., key)

432. Frohne, W.C. 1955b. Ecological by-lines of an Alaskan mosquito worker. Proc. 23rd Calif. Mosquito Control Assoc. Conf.: 98-101. (6 refs.)

433. Frohne, W.C. 1956. The biology of northern mosquitoes. Publ. Hlth Rep. 71(6): 616-621. (15 refs.)

434. Gagné, R.J. 1975. A revision of the nearctic species of the genus Phronia (Diptera: Mycetophilidae). Trans. Am. ent. Soc. 101: (2): 227-318. (90 refs., 35 maps + 141 figs., key)

435. Galloway, T.D. and R.A. Brust. 1976. Observations on mermithid parasites of mosquitoes in Manitoba, Canada. Proc. int. Colloq. Invertebr. Pathol. 1: 227-231. (text refs., 4 figs.)

436. Galloway, T.D. and R.A. Brust. 1979. Review of the genus Romanomermis (Nematoda: Mermithidae) with a description of R. communensis sp.n. from Canada. Can. J. Zool. 57(2): 281-289. (19 refs., 4 figs., 1 tab.)

437. Geary, J.M. 1953. A survey of invertebrate disease vectors in Alaska. Arct. Aeromed. Lab., Ladd Air Force Base, Alaska, Proj. 22-1401-0004, Res. Rep. 1. [not seen] [see 445]

438. Gerstacker, A. and A. Pansch. 1874. Hymenopteren und Dipteren. pp. 404-406 in Geographische Gesellschaft in Bremen. Die zweite deutsche Nordpolarfahrt in den Jahren 1869 und 1870. Bd. 2. (text refs.)

439. Gertsch, W.J. 1933. New genera and species of North American spiders. Am. Mus. Novit. 636. 28 pp. (text refs., 49 figs., text tabs.)

440. Gertsch, W.J. 1934. Notes on American Lycosidae. Am. Mus. Novit. 693. 25 pp. (text refs., key)

441. Gertsch, W.J. 1939. A revision of the typical crab-spiders (Misumeninae) of America north of Mexico. Bull. Am. Mus. nat. Hist. 76(7): 277-442. (197 refs., 271 figs., text tabs., keys)

442. Gertsch, W.J. 1953. The spider genera Xysticus, Coriarachne, and Oxyptila (Thomisidae, Misumeninae) in North America. Bull. Am. Mus. nat. Hist. 102(4): 413-482. (text refs., 93 figs., keys)

443. Gibson, A. 1920. Part I: Lepidoptera. The Lepidoptera collected by the Canadian Arctic Expedition, 1913-1918. Rep. Can. Arctic Expedition 1913-1918. Vol. III: Insects. 58 pp. (0 ref., 68 figs.)

444. Gill, D. and R.W. Strandtmann. 1977. Ectoparasites of the collared lemming (Dicrostonyx torquatus) on Bathurst Island, N.W.T., Canada. J. med. Ent. 14(1): 101-106. (10 refs., 10 figs.)

445. Gill, G.D. 1954. A survey of invertebrate disease vectors in Alaska. Arctic Aeromedical Laboratory. (Project 22-1401-0004, report no. 2) [Tech. Rept. TR 59-(1954)]. 50 pp. (108 refs., 1 fig., 11 tabs., appendix)

446. Gill, G.D. 1955. Filth flies of Central Alaska. J. econ. Ent. 48(6): 648-653. (9 refs., 1 fig., 4 tabs.)

447. Gill, G.D. 1962. The Heleomyzid flies of America North of Mexico (Diptera: Helomyzidae). Proc. U.S. natn. Mus. 113(3465): 495-603. (132 refs., 96 figs., keys)

448. Gillham, N.W. and P.R. Ehrlich. 1954. The butterfly types of Henry Skinner and co-authors in the Academy of Natural Sciences of Philadelphia. (Lepidoptera: Papilionoidea and Hesperioidea). Trans. Am. ent. Soc. 80(3-4): 91-117. (5 + text refs.)

449. Gjullin, C.M. 1946. A key to the Aedes females of America north of Mexico (Diptera, Culicidae). Proc. ent. Soc. Wash. 48(9): 215-236. (10 + text refs., 2 figs.)

450. Gjullin, C.M., R.I. Sailer, A. Stone, and B.V. Travis. 1961. The mosquitoes of Alaska. U.S. Dep. Agric., Agric. Handbk 182. 98 pp. (116 refs., 12 maps, 81 figs., 1 tab., keys)

451. Godfrey, W.E. 1953. Notes on Ellesmere Island birds. Can. Fld Nat. 67: 89-93. (7 refs.)

452. Goetghebuer, M. 1933. Chironomides du Groenland oriental, du Svalbard et de la terre de François Joseph. (Zoological results of the Norwegian Scientific Expeditions to East Greenland. II). Skr. Svalbard Ishavet 53: 19-31. (text refs., 15 figs., key)

453. Goetghebuer, M. 1955. Deux chironomides nouveaux du Groenland. Bull. Assoc. philo. d'Alsace Lorraine 9(3): 167. (0 ref., 1 fig.)

454. Gorham, J.R. 1972. Studies of the biology and control of arthropods of health significance in Alaska. 4. Ecological studies of biting flies on the north slope of Alaska: 1970. Arctic Health Research Center, Fairbanks, Alaska. 62 pp. (36 refs., 20 figs., 13 tabs., 1 appendix)

455. Gorham, J.R. 1976. Orchid pollination by Aedes mosquitoes in Alaska. Am. midl. Nat. 95(1): 208-210. (10 refs., 1 tab.)

456. Gough, H.J. 1974. The first record of Willemia scandinavica Stach (Collembola, Hypogastruridae) from Greenland. Entomologist's mon. Mag. 110(1322-1324): 172. (0 ref.)

457. Graversen, C.B. 1931. Notizen über Grönländische Oribatiden. Meddr Grønland 91(2). 18 pp. (26 refs., 8 figs., 3 tabs.)

458. Graystone, P. 1959. The biting flies of the Fort Churchill area and methods of control at present in use. Def. Res. Bd Can. Northern Lab. Tech. Note 1. 5 pp. (0 ref., 1 fig., 1 appendix)

459. Green, G.H. and J.J.D. Greenwood (Eds.). 1977. Report of the Joint Biological Expedition to N.E. Greenland 1974. Dundee: Dundee University N.E. Greenland Expedition. 258 pp. (299 refs., figs., tabs.)

460. Green, G.H., J.J.D. Greenwood, and C.S. Lloyd. 1977a. The effect of

snow conditions on the date of breeding in waders. pp. 75-78 in G.H. Green and J.J.D. Greenwood (Eds.), Report of the Joint Biological Expedition to N.E. Greenland 1974. Dundee: Dundee University N.E. Greenland Expedition. 258 pp. [précis of Green et al. 1977b] (refs., 1 fig., 6 tabs.)

461. Green, G.H., J.J.D. Greenwood, and C.S. Lloyd. 1977b. The influence of snow conditions on the date of breeding of wading birds in northeast Greenland. J. Zool. 183(3): 311-328. (49 refs., 4 figs., 6 tabs.)

462. Greenwood, J.J.D. 1974a. The distribution and activity of tundra arthropods. pp. 48-55 in R.M.G. O'Brien and J.J.D. Greenwood (Eds.), Report of the University of Dundee N.E. Greenland Expedition 1972. Dundee: Dundee University N.E. Greenland Expedition. 98 pp. (6 refs., 2 figs., 2 tabs.)

463. Greenwood, J.J.D. 1974b. Energy content of some tundra arthropods and plant propagules. p. 56 in R.M.G. O'Brien and J.J.D. Greenwood (Eds.), Report of the University of Dundee N.E. Greenland Expedition 1972. Dundee: Dundee University N.E. Greenland Expedition. 98 pp. (0 ref., 1 tab.)

464. Greenwood, J.J.D. 1974c. General zoological notes. p. 85 in R.M.G. O'Brien and J.J.D. Greenwood (Eds.), Report of the University of Dundee N.E. Greenland Expedition 1972. Dundee: Dundee University N.E. Greenland Expedition. 98 pp. (0 ref.)

465. Greenwood, J.J.D. 1977. Activity patterns and habitat distribution of tundra invertebrates. pp. 168-182 in G.H. Green and J.J.D. Greenwood (Eds.), Report of the Joint Biological Expedition to N.E. Greenland 1974. Dundee: Dundee University N.E. Greenland Expedition. 258 pp. (refs., 6 figs., 10 tabs.)

466. Greenwood, J.J.D., S.W.G. Enticott, R.A. Herbert, and G. Lochhead. 1977. Biological studies of tarns near Mestersvig. pp. 183-211 in G.H. Green and J.J.D. Greenwood (Eds.), Report of the Joint Biological Expedition to N.E. Greenland 1974. Dundee: Dundee University N.E. Greenland Expedition. 258 pp. (refs., 19 figs., 13 tabs.)

467. Gregg, R.E. 1972. The northward distribution of ants in North America. Can. Ent. 104(7): 1073-1091. (14 refs., 10 figs., 1 tab.)

468. Greiner, E.C., G.F. Bennett, E.M. White, and R.F. Coombs. 1975. Distribution of the avian hematozoa of North America. Can. J. Zool. 53(12): 1762-1787. (139 refs., 5 figs., 1 tab.)

469. Gressitt, J.L. and C.M. Yoshimoto. 1974. Insect dispersal studies in northern Alaska. Pacif. Insects 16(1): 11-30. (16 refs., 2 + 7 figs., 10 tabs.)

470. Griffiths, G.C.D. 1966. The Agromyzidae (Diptera) of Greenland. Meddr Grønland 170(4). 32 pp. (21 refs., 21 figs.)

471. Griffiths, G.C.D. 1973. Studies on boreal Agromyzidae (Diptera). IV. Phytomyza miners on Angelica, Heracleum, Laserpitium and Pastinaca (Umbelliferae). Quaest. ent. 9(3): 219-253. (11 refs., 14 figs., key)

472. Griffiths, G.C.D. 1976. Studies on boreal Agromyzidae (Diptera). XII. Phytomyza and Chromatomyia miners on Astereae (Compositae).

Quaest. ent. 12(3): 239-278. (98 refs., 51 figs., key)

473. Gunder, J.D. 1932. New Rhopalocera (Lepidoptera). Can. Ent. 64(12): 276-284. (0 ref.)

474. Gurney, A.B. and A.R. Brooks. 1959. Grasshoppers of the mexicanus group, genus Melanoplus (Orthoptera: Acrididae). Proc. U.S. natn. Mus. 110(3416): 1-93, 5 pls. (180 refs., 18 figs., 5 pls., 1 tab., key)

475. Haarløv, N. 1942. A morphologic-systematic-ecological investigation of Acarina. Meddr Grønland 128(1). 71 pp., pls. 1-3. (84 refs., 26 figs., 3 pls., 1 tab., key)

476. Haarløv, N. 1967. Arthropoda (Acarina, Diptera) from subfossil layers in West Greenland. Meddr Grønland 184(3). 17 pp., pls. I-VIII. (32 refs., 2 figs., 8 pls., 1 tab.)

477. Habeeb, H. 1957. Some water mites from Baffin Island. Leafl. Acadian Biol. No. 13. 2 pp. (0 ref., 6 figs.)

478. Hackman, W. 1959. On the genus Scaptomyza Hardy (Dipt., Drosophilidae) with descriptions of new species from various parts of the world. Acta zool. fenn. 97. 73 pp. (53 refs., 74 figs., keys)

479. Hadwen, S. 1927. Notes on the life history of Oedemagena tarandi L. and Cephenomyia trompe Modeer. J. Parasit. 13(1): 56-65. (5 refs., 2 figs.)

480. Hagen, H.A. 1861. Synopsis of the Neuroptera of North America, with a list of the South American species. Smithson. misc. Collns 4(2). 368 pp. (1-20; 1-347). (93 refs., 3 tabs.)

481. Halašková, V. 1977. A revision of the genera of the family Zerconidae (Acari: Gamasides) and descriptions of new taxa from several areas of nearctic region. Stud. Československ. Akad. Ved. No. 7. 74 pp. (38 refs., 19 figs., text tabs., keys)

482. Hall, D.G. 1948. The blowflies of North America. In Entomological Society of America. Thomas Say Foundation, Washington, D.C., 477 pp. (303 refs., 46 pls., 5 colour pls., 6 figs., keys)

483. Hamilton, J. 1894. Catalogue of the Coleoptera of Alaska, with the synonymy and distribution. Trans. Am. ent. Soc. 21: 1-38. (text refs., text tabs.)

484. Hamilton, K.G.A. 1975. A review of the northern hemisphere Aphrodina (Rhynchota: Homoptera: Cicadellidae), with special reference to the nearctic fauna. Can. Ent. 107(10): 1009-1027. (114 refs., 22 figs., 1 tab., keys)

485. Hamilton, K.G.A. and H.H. Ross. 1975. New species of grass-feeding Deltocephaline leafhoppers with keys to the nearctic species of Palus and Rosenus (Rhynchota: Homoptera: Cicadellidae). Can. Ent. 107(6): 601-611. (3 refs., 59 figs., keys)

486. Hammer, M. 1937. A quantitative and qualitative investigation of the microfauna communities of the soil at Angmagssalik and in Mikis Fjord. Meddr Grønland 108(2). 53 pp. (33 refs., 17 figs., 15 tabs.)

487. Hammer, M. 1938a. **The zoology of East Greenland. Collemboles.**

Meddr Grønland 121(2). 42 pp. (72 refs., 2 figs., 3 tabs.)

488. Hammer, M. 1938b. Eine neue Isotoma-Art aus Ostgrönland. Zool. Anz. 121 (1-2): 43-45. (3 refs., 2 figs.)

489. Hammer, M. 1944. Studies on the Oribatids and Collemboles of Greenland. Meddr Grønland 141(3). 210 pp. (154 refs., 38 figs., 67 tabs.)

490. Hammer, M. 1946. The zoology of East Greenland. Oribatids. Meddr Grønland 122(1). 39 pp. (42 refs., 4 figs., 3 tabs.)

491. Hammer, M. 1952a. Investigations on the microfauna of northern Canada, Part I. Oribatei. Acta arct. 4. 108 pp. (64 refs., 106 + 13 figs., 6 tabs.)

492. Hammer, M. 1952b. The Oribatid and Collembola fauna in some soil samples from Søndre Strømfjord. Ent. Meddr 26(5): 404-414. (10 refs., 3 figs., 1 tab.)

493. Hammer, M. 1953a. Investigations on the microfauna of northern Canada, part 2. Collembola. Acta arct. 6. 108 pp. (99 refs., 10 figs., 26 pls., 5+ tabs.)

494. Hammer, M. 1953b. Collemboles and oribatids from the Thule District, North West Greenland and Ellesmere Island, Canada; collected by J.C. Troelsen and Chr. Vibe. Meddr Grønland 135(5). 16 pp. (16 refs., 15 figs.)

495. Hammer, M. 1954. Collemboles and oribatids from Peary-land (North Greenland). Meddr Grønland 127(5). 28 pp. (20 refs., 10 figs., 2 tabs.)

496. Hammer, M. 1955a. Alaskan Oribatids. Acta arct. 7. 36 pp. (30 refs., 23 figs.)

497. Hammer, M. 1955b. Some aspects of the distribution of microfauna in the arctic. Arctic 8(2): 115-126. (11 refs., 12 figs.)

498. Hammer, M. 1960. A few more Oribatids from Greenland and Lapland. Ent. Meddr 29(7): 360-363. (18 refs.)

499. Hanbury, D.T. 1904. Sport and travel in the northland of Canada. MacMillan Co., N.Y. 319 pp. [see Elwes 1904]

500. Hanna, G.D., C.P. Alexander, F.R. Cole, T.H. Frison, A.D. MacGillivray, J.R. Malloch, E.P. Van Duzee, M.C. Van Duzee, and E.C. Van Dyke. 1921. Insects of the Pribilof Islands, Alaska. Proc. Calif. Acad. Sci. ser. 4, 11(14): 153-195. [see authors]

501. Hansen, D.C. and E.F. Cook. 1976. The systematics and morphology of the nearctic species of Diamesa Meigen, 1835 (Diptera: Chironomidae). Mem. Am. ent. Soc. 30. 203 pp. (236 refs., 149 figs., text + 1 tab.)

502. Hardwick, D.F. 1966. A synopsis of the westermanni group of the genus Euxoa Hbn. (Lepidoptera: Noctuidae) with descriptions of two new species. Can. Ent. 98(7): 760-768. (text refs., 12 figs., key)

503. Hardwick, D.F. 1970. The genus Euxoa (Lepidoptera: Noctuidae) in North America. 1. Subgenera Orosagrotis, Longivesica, Chorizagrotis, Pleonectopoda, and Crassivesica. Mem. ent. Soc.

Can. 67. 177 pp. (159 refs., 326 figs., keys)

504. Hardy, D.E. 1945. Revision of nearctic Bibionidae including neotropical Plecia and Penthetria (Diptera). Univ. Kans Sci. Bull. 30(II, 15): 367-547. (text refs., 219 figs., keys)

505. Harper, F. 1955. The barren ground caribou of Keewatin. Univ. Kans. Publs Mus. nat. Hist. 6. 164 pp. (229 refs., 28 figs., 1 map, text tabs.)

506. Harper, F. 1956. The mammals of Keewatin. Univ. Kans. Publs Mus. nat. Hist. 12. 94 pp., pl. 1-6. (99 refs., 8 figs., 6 pls., 1 map)

507. Harper, F. 1961. Land and freshwater mammals of the Ungava Peninsula. Univ. Kans. Publs Mus. nat. Hist. 27. 178 pp. (252 refs., 3 figs., 45 maps, 8 pls., 2 tabs.)

508. Harper, P.P., J.G. Pilon, and J.M. Perron. 1975. Insectes aquatiques du nord du Québec. Ann. Soc. ent. Queb. 20(1): 33-43. (19 refs., 1 fig.)

509. Hatch, M.H. 1939. A key to the species of Nebria of northwestern North America (Coleoptera, Carabidae). Pan-Pacif. Ent. 15(3): 117-122. (16 refs., key)

510. Haufe, W.O. 1953. Predicting mosquito emergence. Proc. 40th A. Mtg. N.J. Mosquito Exterm. Ass.: 52-57. (5 refs., 3 figs.)

511. Haufe, W.O. 1957. Physical environment and behaviour of immature stages of Aedes communis (Deg.) (Diptera: Culicidae) in subarctic Canada. Can. Ent. 89(2): 120-139. (16 refs., 8 figs., 4 tabs.)

512. Haufe, W.O. 1966a. Synoptic correlation of weather with mosquito activity. Biometeorology 2. (Proc. 3rd int. Congr. Biometeor. 1963): 523-540. (16 refs., 12 figs.)

513. Haufe, W.O. 1966b. The significance of biometeorology in the ecology of insects. Int. J. Bioclim. Biomet. 10(3): 241-252. (60 refs., 4 figs.)

514. Haufe, W.O. and L. Burgess. 1956. Development of Aedes (Diptera: Culicidae) at Fort Churchill, Manitoba, and prediction of dates of emergence. Ecology 37(3): 500-519. (36 refs., 8 figs., 5 tabs.)

515. Hayward, R. 1908. Studies in Amara. Trans. Am. ent. Soc. 34(1): 13-65. (text refs., key)

516. Hebard, M. 1923. Orthoptera. p. 140 in W.L. McAtee (Ed.), A biological survey of the Pribilof Islands, Alaska. Part II. Insects, arachnids and chilipods of the Pribilof Islands, Alaska. N. Am. Fauna 46: 129-255. (text refs.)

517. Hebard, M. 1935. Notes on the group Gomphoceri and a key to its genera, including one new genus (Orthoptera, Acrididae, Acridinae). Ent. News 46(7-8): 184-188; 204-208. (text refs.)

518. Hebard, M. 1936. Notes on North American Orthoptera of the arctic-alpine zone. Ent. News 47(1): 13-15. (text refs.)

519. Hedeen, R.A. and H.L. Keegan. 1952. The use of Lindane and Dieldrin

as mosquito adulticides in Alaska. Mosquito News 12(4): 242-243. (1 ref.)

520. Heie, O.E. 1979. Revision of the aphid genus Nasonovia Nordvilko, including Kakimia Hottes and Frison, with keys and descriptions of the species of the world (Homoptera, Aphididae). Ent. Scand. Suppl. 9. 105 pp. (116 refs., 96 figs., 14 tabs., keys)

521. Heinrich, C. 1935. Heterocera. The exploration of Southampton Island, Hudson Bay, by G.M. Sutton. Mem. Carnegie Mus. 12, part II, sect. 5: 27-34. (text refs., 1 fig., 1 pl.)

522. Heinrich, C. 1956. American moths of the subfamily Phycitinae. Bull. U.S. natn. Mus. 207: I-VIII, 1-581. (text refs., 1138 figs., 1 tab., keys)

523. Heinrich, G.H. 1956a. A report on some North American arctic and subarctic Ichneumoninae. Can. Ent. 88(8): 477-487. (text refs., key)

524. Heinrich, G.H. 1956b. Ichneumon lariae Curtis. A critical study on this arctic species and its closely related forms. Can. Ent. 88(12): 686-691. (text refs., 1 tab.)

525. Heinrich, G.H. 1956c. Holarctic elements among the Ichneumoninae of Canada. Can. Ent. 88(11): 647-652. (text refs., key)

526. Heinrich, G.H. 1958. A report on some North American arctic and sub-arctic Ichneumoninae. First supplement. Can. Ent. 90(12): 739-741. (0 ref.)

527. Heinrich, G.H. 1961a. Synopsis of Nearctic Ichneumoninae Stenopneusticae with particular reference to the northeastern region (Hymenoptera). Part I. Can. Ent. Suppl. 15 [1960]. pp. 3-87. (text refs., 30 figs., 4 tabs., keys)

528. Heinrich, G.H. 1961b. Synopsis of Nearctic Ichneumoninae Stenopneusticae with particular reference to the northeastern region (Hymenoptera). Part II. Can. Ent. Suppl. 18 [1960]. pp. 91-205. (text refs., 54 figs., 3 tabs., keys)

529. Heinrich, G.H. 1961c. Synopsis of nearctic Ichneumoninae Stenopneusticae with particular reference to the northeastern region (Hymenoptera). Part III. Can. Ent. Suppl. 21. pp. 209-368. (text refs., 73 figs., 5 tabs., keys)

530. Heinrich, G.H. 1961d. Synopsis of Nearctic Ichneumoninae Stenopneusticae with particular reference to the northeastern region (Hymenoptera). Part IV. Can. Ent. Suppl. 23. pp. 371-505. (text refs., 65 figs., 4 tabs., keys)

531. Heinrich, G.H. 1962a. Synopsis of Nearctic Ichneumoninae Stenopneusticae with particular reference to the northeastern region (Hymenoptera). Part V. Can. Ent. Suppl. 26. pp. 509-671. (text refs., 50 figs., 3 tabs., keys)

532. Heinrich, G.H. 1962b. Synopsis of nearctic Ichneumoninae Stenopneusticae with particular reference to the northeastern region (Hymenoptera). Part VII. Can. Ent. Suppl. 29. pp. 805-886. Addenda and Corrigenda pp. 861-884. (text refs.)

533. Hemmingsen, A.M. 1956. Deep-boring ovipository instincts of some crane-fly species (Tipulidae) of the subgenera Vestiplex Bezzi

and Oreomyza Pok., and some associated phenomena. Vidensk. meddr. dansk naturh.Foren. 118: 243-315. (89 refs., 16 figs., 5 tabs)

534. Hemmingsen, A.M. and B. Jensen. 1957. The occurrence of Tipula (Vestiplex) arctica Curtis in Greenland and its decreasing body length with increasing latitude. Meddr Grønland 159(1). 20 pp. (refs., 4 figs., 2 pls.)

535. Henriksen, K.L. 1924. [The Canadian Arctic Expedition 1913-1918.] Ent. Meddr 14(6): 235-239. (title in English, text in Danish) (1 text ref.)

536. Henriksen, K.L. 1935. Insects and Arachnids I. Insects (exclusive of Diptera) and Acarina. The Scoresby Sound Committee's 2nd East Greenland Expedition in 1932 to King Christian IX's land. Meddr Grønland 104(16): 1-9. (13 refs.)

537. Henriksen, K.L. 1937. Insects collected on the Fifth Thule expedition. Rep. Fifth Thule exped. 1921-1924 2(8). 34 pp.(text refs., 1 map)

538. Henriksen, K.L. 1939. A revised index of the insects of Grønland, containing a supplement to the insect list in Kai L. Henriksen & Will Lundbeck: Grønlands landarthropoder (Medd. om Grønl. 22. 1917). Meddr Grønland 119(10). 112 pp. (78 refs., index)

539. Henriksen, K.L. and W. Lundbeck. 1917. Grønlands Landarthropoder (Insecta et Arachnida). Meddr Grønland 22: 481-821. (In Danish) (82 refs., 4 figs.)

540. Herrin, C.S. 1970. A systematic revision of the genus Hirstionyssus (Acari: Mesostigmata) of the nearctic region. J. med. Ent. 7(4): 391-437. (66 refs., 77 figs., keys)

541. Hewitt, C.G. 1920. Introduction, and list of new genera and species collected by the expedition. Rep. Can. Arctic Expedition 1913-1918. Vol. III: Insects. xi pp.

542. Hewitt, C.G. et al. 1919-1922. Report of the Canadian Arctic Expedition, 1913-1918, Vol.III(Parts A to L); Insects. Ottawa, King's Printer. [see Canadian Arctic Expedition 1913-1918]

543. Hicks, S.D. 1953. Distribution of Carabus vietinghoffii in North America. Coleopts Bull. 7: 1-3. (7 refs., 1 fig.)

544. Hicks, S.C. 1955. Natural history survey of Coppermine, Northwest Territories, 1951. Can. Fld Nat. 69(4): 162-166. (4 refs.)

545. Hille Ris Lambers, D. 1952. The aphid fauna of Greenland. Meddr Grønland 136(1). 33 pp. (0 ref., 25 figs., text tabs.)

546. Hille Ris Lambers, D. 1960. Additions to the aphid fauna of Greenland. Meddr Grønland 159(5). 18 pp. (text refs., text tabs.)

547. Hillyard, P.D. 1979. Notes on two collections of spiders from Baffin Island and Greenland. Bull.Br.Arachnol.Soc. 4(7): 296-297. (2 refs.)

548. Hine, J.S. 1922. Descriptions of Alaskan Diptera of the family Syrphidae. Ohio J. Sci. 22(5): 143-147. (0 ref.)

549. Hine, J.S. 1923. Alaskan species of Diptera of the genus Helophilus,

with notes on others. Ohio J. Sci. 23(4): 192-200. (text refs., key)

550. Hirvenoja, M. 1973. Revision der Gattung Cricotopus van der Wulp and ihrer Verwandten (Diptera, Chironomidae). Ann. zool. fenn. 10(1): 1-363. (658 refs., 216 figs., 127 tabs., keys)

551. Hobbie, J.E. 1975. Some controls in tundra pond ecosystems. pp. III-43-III-51 in Proc. Circumpolar conference on Northern Ecology. N.R.C., Ottawa. (5 refs., 2 figs.)

552. Hocking, B. 1950. Further tests of insecticides against black flies (Diptera: Simuliidae) and a control procedure. Scient. Agric. 30(12): 489-508. (11 refs., 9 figs., 5 tabs., appendix)

553. Hocking, B. 1952a. Protection from northern biting flies. Mosquito News 12(2): 91-102. (73 refs., 2 tabs.)

554. Hocking, B. 1952b. Autolysis of flight muscles in a mosquito. Nature, Lond. 169(4313): 1101. (3 refs.)

555. Hocking, B. 1953a. The intrinsic range and speed of flight of insects. Trans. R. ent. Soc. Lond. 104(8): 223-345. (505 refs., 29 figs., 6 pls., 18 tabs., 3 appendixes)

556. Hocking, B. 1953b. Notes on the activities of Aedes larvae. Mosquito News 13(2): 77-81. (11 refs., 3 figs.)

557. Hocking, B. 1953c. Developments in the chemical control of black flies (Diptera, Simuliidae). Can. J. agric. Sci. 33(6): 572-578. (5 refs., 1 tab.)

558. Hocking, B. 1954. Flight muscle autolysis in Aedes communis (DeGeer). Mosquito News 14(3): 121-123. (7 refs., 3 figs., 1 tab.)

559. Hocking, B. 1960. Northern biting flies. A. Rev. Ent. 5: 135-152. (290 refs.)

560. Hocking, B. 1968. Insect-flower associations in the high arctic with special reference to nectar. Oikos 19(2): 359-387. (22 refs., 6 figs., 8 tabs., 2 appendixes)

561. Hocking, B. and L.R. Pickering. 1954. Observations on the bionomics of some northern species of Simuliidae (Diptera). Can. J. Zool. 32(2): 99-119. (17 refs., 1 fig., 5 tabs.)

562. Hocking, B., W.R. Richards, and C.R. Twinn. 1950. Observations on the bionomics of some northern mosquito species (Culicidae: Diptera). Can. J. Res. (D) 28(2): 58-80. (17 refs., 17 figs., 4 tabs.)

563. Hocking, B. and C.D. Sharplin. 1964. Bees at 82°N. Bee World 45(4): 144-146. (5 refs., 1 fig.)

564. Hocking, B. and C.D. Sharplin. 1965. Flower basking by arctic insects. Nature, Lond. 206(4980): 215. (1 ref.)

565. Hocking, B., C.R. Twinn, and W.C. McDuffie. 1949. A preliminary evaluation of some insecticides against immature stages of black flies (Diptera: Simuliidae). Scient. Agric. 29(2): 69-80. (8 refs., 10 figs., 1 tab.)

566. Hodkinson, I.D. 1978. The psyllids (Homoptera: Psylloidea) of Alaska. Syst. Ent. 3(4): 333-360. (28 refs., 95 figs., 1 tab., keys)

567. Hodkinson, I.D., T.S. Jensen, and S.F. MacLean, Jr. 1979. The distribution, abundance and host plant relationship of Salix-feeding psyllids (Homoptera: Psylloidea) in arctic Alaska. Ecol. Ent. 4(2): 119-132. (21 refs., 6 figs., 6 tabs.)

568. Høeg, O.A. 1929. Pollen on bumble-bees from Ellesmere Land. K. norske Vidensk. Selsk. Forh. 2(16): 55-57. (0 ref.)

569. Holding, A.J., O.W. Heal, S.F. MacLean, and P.W. Flanagan (Eds.). 1974. Soil organisms and decomposition in tundra. Tundra Biome Steering Committee, Stockholm. 398 pp. [see MacLean 1974; Whittaker 1974]

570. Holland, G.P. 1944. Notes on some northern Canadian Siphonaptera, with a description of a new species. Can. Ent. 76(12): 242-246. (text + 1 ref., 10 figs.)

571. Holland, G.P. 1949. [reprinted 1968] The Siphonaptera of Canada. Can. Dep. Agric. Publ. 817. Tech. Bull. 70. 306 pp. (211 refs., 350 figs., 44 maps, 4 tabs., keys)

572. Holland, G.P. 1950. Notes on Megabothris asio (Baker) and M. calcarifer (Wagner) with the description of a new subspecies (Siphonaptera: Ceratophyllidae). Can. Ent. 82(6): 126-133. (text refs., 8 figs.)

573. Holland, G.P. 1951. Notes on some bird fleas, with the description of a new species of Ceratophyllus, and a key to the bird fleas known from Canada (Siphonaptera: Ceratophyllidae). Can. Ent. 83(11): 281-289. (12 refs., 10 figs.)

574. Holland, G.P. 1952a. Notes on Siphonaptera from Canada. Can. Ent. 84(3): 65-73. (27 refs., 3 figs.)

575. Holland, G.P. 1952b. Descriptions of fleas from northern Canada (Siphonaptera). Can. Ent. 84(10): 297-308. (10 refs., 25 figs.)

576. Holland, G.P. 1958. Distribution patterns of northern fleas (Siphonaptera). Proc. 10th int. Congr. Ent. (Montreal 1956) 1: 645-658. (20 refs., 6 figs.)

577. Holland, G.P. 1962. Summary of entomological activities in the Arctic and sub-Arctic by the Canada Department of Agriculture. 8 pp, in WHO Conference on medicine and public health in the arctic and antarctic. World Health organization Arctic Conf. document no. 34. (text refs., 1 fig.)

578. Holland, G.P. 1963. Faunal affinities of the fleas (Siphonaptera) of Alaska. pp. 45-63 in J.L. Gressitt (Ed.), Pacific Basin biogeography. Bishop Museum Press, Honolulu (Symp. 10th Pacific Sci. Congr. 1961). 563 pp. (4 refs., 9 figs.)

579. Holland, G.P. and W.L. Jellison. 1952. On Mioctenopsylla Rothschild, a genus of Siphonaptera (Ceratophyllidae) with a description of a new species. Can. Ent. 84(12): 374-379. (6 refs., 10 figs.)

580. Holland, W.J. (Ed.). 1932-1935. The exploration of Southampton

Island, Hudson Bay, by George Miksch Sutton. Mem. Carnegie Mus. 12, part I, sections 1-3; Part II, sections 1-5. [see Brooks et al. 1935; Crosby et al. 1934]

581. Holland, W.J. and A. Avinoff. 1935. Rhopalocera. II. The Lepidoptera collected by G.M. Sutton on Southampton Island. The exploration of Southampton Island, Hudson Bay, by G.M. Sutton. Mem. Carnegie Mus. 12, part II, sect. 5: 3-26. (text refs., 3 pls.)

582. Holm, A. 1937. Notes on the spider fauna of East Greenland. Ark. Zool. 29B(12): 2 pp. (0 ref.)

583. Holm, A. 1939. Insecta and Araneae collected in the Kangerdlugsuak region of East Greenland by the British expedition, 1935-36. Part II. Araneae. Ann. Mag. nat. Hist. 11(3): 72-80. [see Carpenter and Holm 1939]

584. Holm, A. 1956. Notes on arctic spiders of the genera Erigone Aud. and Hilaira Sim. Ark. Zool. ser 2, 9(20): 453-468. (27 refs., 5 figs., 2 tabs.)

585. Holm, A. 1958. Spiders (Araneae) from Greenland. Ark. Zool., ser. 2, 11(31): 525-534. (13 refs., 10 figs.)

586. Holm, A. 1960. On a collection of spiders from Alaska. Zool. Bidr. Upps. 33. 132 pp., IV pls. (19 refs., 4 + 40 figs.)

587. Holm, A. 1961. Notes on arctic spiders. Ark. Zool.,ser. 2, 12(32): 511-514. (13 refs., 3 figs.)

588. Holm, A. 1967. Spiders (Araneae) from West Greenland. Meddr Grønland 184(1). 99 pp. (70 refs., 103 figs., 1 tab.)

589. Holm, A. 1970. Notes on spiders collected by the "Vega" expedition 1878-1880. Ent. Scand. 1(3): 188-208. (30 + text refs., 42 figs.)

590. Holm, A. 1973. On the spiders collected during the Swedish expeditions to Novaya Zemlya and Yenisey in 1875 and 1876. Zool. Scripta 2(2-3): 71-110. (84 refs., 117 figs., text tabs.)

591. Holmes, R.T. 1963. Food overlap in breeding sandpipers in Northern Alaska. Proc. 14th Alaska Sci. Conf.: 66-67. [abstract only]

592. Holmes, R.T. 1964. Breeding of arctic sandpipers in relation to food supply. Bull. ecol. Soc. Am. 45: 114. [abstract only]

593. Holmes, R.T. 1966a. Breeding ecology and annual cycle adaptations of the red-backed sandpiper (Calidris alpina) in northern Alaska. Condor 68(1): 3-46. (72 refs., 15 figs., 2 tabs.)

594. Holmes, R.T. 1966b. Feeding ecology of the red-backed sandpiper (Calidris alpina) in arctic Alaska. Ecology 47(1): 32-45. (22 refs., 7 figs., 3 tabs.)

595. Holmes, R.T. 1970. Differences in population density, territoriality, and food supply of dunlin on arctic and subarctic tundra. pp. 303-319 in A. Watson (Ed.), Animal populations in relation to their food resources. (Symp. Br. Ecol. Soc. 1969). Blackwell, Oxford. 477 pp. (16 refs., 1 fig., 4 tabs.)

596. Holmes, R.T. and F.A. Pitelka. 1968. Food overlap among coexisting

sandpipers on northern Alaskan tundra. Syst. Zool. 17(3): 305-318. (16 refs., 8 figs., 2 tabs.)

597. Holmgren, A.E. 1872. [Insects from North Greenland collected by Prof. A.E. Nordenskiöld in 1870.] Öfvers. K. VetenskAkad. Förh. 29(6): 97-105. (In Swedish) (text refs.)

598. Holmquist, C. 1975. Lakes of northern Alaska and northwestern Canada and their invertebrate fauna. Zool. Jb. Syst. 102(3): 333-484. (274 refs., 58 figs., 4 tabs.)

599. Homeyer, A. 1874. Lepidopteren. pp. 407-410 in Geographische Gesellschaft in Bremen. Die zweite deutsche Nord-polarfahrt in den Jahren 1869 und 1870. Bd. 2.

600. Hopkins, D.M., R.E. Giterman, and J.V. Matthews, Jr. 1976. Interstadial mammoth remains and associated pollen and insect fossils, Kotzebue Sound area, northwestern Alaska. Geol. 4: 169-172. (21 refs., 2 figs., 2 tabs.)

601. Hopkins, D.M., J.V. Matthews, Jr., J.A. Wolfe, and M.L. Silberman. 1971. A Pliocene flora and insect fauna from the Bering Strait region. Palaeogeogr., Palaeoclimatol., Palaeoecol. 9(3): 211-231. (29 refs., 4 figs., 5 tabs.)

602. Hopla, C.E. 1964-1965. The feeding habits of Alaskan mosquitoes Bull. Brooklyn ent. Soc. 59/60: 88-127. (86 refs., 9 figs., 7 tabs.)

603. Hopla, C.E. 1965a. Alaskan hematophagous insects, their feeding habits and potential as vectors of pathogenic organisms; I: The Siphonaptera of Alaska. Tech. Rep. U.S. Air Force Arct. Aeromed. Lab. TR-64-12, Vol. I. 267 pp. (155 refs., 6 figs., 14 maps, 33 pls., 17 tabs.)

604. Hopla, C.E. 1965b. Alaskan hematophagous insects, their feeding habits and potential as vectors of pathogenic organisms, II. The feeding habits and colonization of subarctic mosquitoes. Tech. Rep. U.S. Air Force Arct. Aeromed. Lab. TR-64-12, Vol. II. 90 pp. (148 refs., + bibliography (71 refs.) on cold hardiness, 34 figs., 3 tabs.)

605. Horn, G.H. 1876. Synonymy of the Coleoptera of the fauna Boreali-Americana, Kirby. Can. Ent. 8: 126-130, 150-151, 166-170, 190-193. (0 ref.)

606. Horsfall, W.R. 1974. Heteromorphic development of Aedine mosquitoes reared at abnormal temperatures. Ann. zool. fenn. 11(3): 224-236. (28 refs., 15 figs.)

607. Horsfall, W.R. and F.R. Voorhees. 1972. Eggs of floodwater mosquitoes XIV. Northern Aedes (Diptera: Culicidae). Ann. ent. Soc. Am. 65(1): 123-126. (2 refs., 12 figs.)

608. Hough, G. deN. 1898a. Two new American species of Cynomyia, a study in chaetotaxy. Ent. News 9(5): 105-111. (0 ref., 2 figs.)

609. Hough, G. deN. 1898b. A third American species of Cynomyia. Ent. News 9(7): 165-166. (0 ref., 1 fig.)

610. [Hovanitz, W.] 1948. Colias in Canada. Lepid. News 2(2): 13. (0 ref.)

611. Hovanitz, W. 1950a. The biology of Colias butterflies, I. The distribution of the North American species. Wasmann J. Biol. 8(1): 49-75. (0 ref., 8 figs.)

612. Hovanitz, W. 1950b. The biology of Colias butterflies, II. Parallel geographical variation of dimorphic color phases in North American species. Wasmann J. Biol. 8(2): 197-219. (0 ref., 6 figs., 9 tabs.)

613. Hovanitz, W. 1951. The biology of Colias butterflies, III. Variation of adult flight in the arctic and subarctic. Wasmann J. Biol. 9(1): 1-9. (0 ref., 1 fig.)

614. Hovanitz, W. 1955. Colias nastes and Colias hecla from the Meade River, Alaska. Wassmann J. Biol. 13(1): 1-8. (0 ref., 2 tabs., 2 figs.)

615. Howe, W.H. et al. 1975. The butterflies of North America. Doubleday, Garden City, N.Y. 633 pp. (116 refs., 31 figs., 97 colour pls.)

616. Huckett, H.C. 1939. Descriptions of new North American Anthomyiidae belonging to the genus Pegomyia Rob.-Desv.(Diptera). Trans. Am. ent. Soc. 65(1): 1-36. (text refs., 21 figs.)

617. Huckett, H.C. 1941. A revision of the North American species belonging to the genus Pegomyia (Diptera: Muscidae). Mem. Am. ent. Soc. 10. 131 pp. (text refs., 94 figs., keys)

618. Huckett, H.C. 1944. A revision of the North American species belonging to the genus Hydrophoria Robineau-Desvoidy (Diptera: Muscidae). Ann. ent. Soc. Am. 37(2): 261-297. (text refs., 51 figs., key)

619. Huckett, H.C. (circa 1958). The Muscidae of the arctic and subarctic regions of North America and of Alaska. Final Report, Arctic Inst. N. Am., project ONR-80. 302 pp. [superseded by Huckett 1965]

620. Huckett, H.C. 1965. The Muscidae of northern Canada, Alaska and Greenland (Diptera). Mem. ent. Soc. Can. 42. 369 pp. (115 refs., 7 + 200 figs., 1 tab., keys)

621. Hudson, J.E. 1977. Alleged "reluctance" of boreal Aedes and Culiseta to take more than one blood meal. Mosquito News 37(1): 140-141. (6 refs., 1 tab.)

622. Hungerford, H.B. 1948. The Corixidae of the Western Hemisphere (Hemiptera). Univ. Kans. Sci. Bull. 32. 827 pp. [reprinted 1977] (text + 40 refs., 112 pls.,keys). [+ 58 refs. in reprint ed.]

623. Hunter, P.E. and R.W. Husband. 1973. Pneumolaelaps (Acarina) mites from North America and Greenland. Fla Ent. 56(2): 77-91. (15 refs., 4 figs., 1 tab.)

624. Hurd, P.D., Jr. 1954. "Myiasis" resulting from the use of the aspirator method in the collection of insects. Science, N.Y. 119: 814-815. (0 ref.)

625. Hurd, P.D., Jr. 1955. [The nature and extent of the insect fauna of northern Alaska.] Pan-Pacif. Ent. 31(1): 36-37. (0 ref.)

626. Hurd, P.D., Jr., and E.E. Lindquist. 1958. Analysis of soil

invertebrate samples from Barrow, Alaska. Final Rep. Arctic Inst. N. Am. project ONR-173 and ONR-193. 24 pp. (mimeo) (3 text refs., 4 appendixes)

627. Hurd, P.D., Jr., and F.A. Pitelka. 1955. The role of insects in the economy of certain arctic Alaskan birds. Proc. 3rd Alaska Sci. Conf.: 136-137. [abstract only]

628. Hurley, R. 1965. A revision of the nearctic species of Hydrophorus (Diptera, Dolichopodidae). Unpub. Ph.D. thesis, Univ. Illinois. 187 pp.

629. Hussell, D.J.T. 1972. Factors affecting clutch size in arctic passerines. Ecol. Monogr. 42(3): 317-364. (102 refs., 18 figs., 23 tabs.)

630. Ide, F.P. 1937. Descriptions of eastern North American species of baetine mayflies with particular reference to the nymphal stages. Can. Ent. 69(10): 219-231. (0 ref., 13 figs.)

631. Ide, F.P., C.R. Twinn, and D.M. Davies. 1958. Seasonal emergence of black flies from streams in northern Canada. Proc. 10th int. Congr. Ent. (Montreal 1956) 3: 809. [abstract only]

632. Ivie, W. 1965. The spiders of the genus Islandiana (Linyphiidae, Erigoninae). Am. Mus. Novit. 2221. 25 pp. (43 refs., 53 figs., key)

633. Ivie, W. 1969. North American spiders of the genus Bathyphantes (Araneae, Linyphiidae). Am. Mus. Novit. 2364. 70 pp. (98 refs., 121 figs., keys)

634. Jachowski, L.A., Jr., and C. Shultz. 1948a. Notes on the biology and control of mosquitoes at Umiat, Alaska. Proj. NM 005 017, Naval Med. Res. Inst. Rep. 1. 17 pp. [see Jachowski and Schultz 1948b]

635. Jachowski, L.A., Jr., and C. Schultz. 1948b. Notes on the biology and control of mosquitoes at Umiat, Alaska. Mosquito News 8(4): 155-165. (2 refs., 9 figs., 2 tabs.)

636. Jackson, A.R. 1930. Results of the Oxford University Expedition to Greenland, 1928. Araneae and Opiliones collected by Major R.W.G. Hingston; with notes on Icelandic spiders. Ann. Mag. nat. Hist. 10(6): 639-656. (38 refs.)

637. Jackson, A.R. 1933. Results of the Oxford University Expedition to Akpatok in 1931: Araneae. Proc. zool. Soc. Lond. (B) 102: 145-159. (30 refs., 10 figs., 1 tab.)

638. Jackson, A.R. 1934. Notes on arctic spiders obtained in 1933. Ann. Mag. nat. Hist. 10(14): 611-620. (11 refs., 7 figs.)

639. Jackson, A.R. 1938. Notes on arctic spiders obtained in 1933-1936. Proc. zool. Soc. Lond. (B) 107: 543-551. (8 refs.)

640. James, H.G. 1938. Notes on some arctic Collembola. Can. Ent. 70(7): 151-154. (4 refs., 11 figs.)

641. James, H.G. 1950. Notes on a preliminary survey of natural control of biting flies (Diptera: Culicidae, Tabanidae) at Churchill, Manitoba. 32nd Rep. Quebec Soc. Prot. Plants: 119-121. (3 refs.)

642. James, H.G. 1952. Natural control of Tabanidae (Diptera) in the region of Churchill, Manitoba. 82nd A. Rep. ent. Soc. Ont. 70-74. (3 refs., 1 fig., 1 tab.)

643. James, H.G. and B.C. Smith. 1958. Observations on three species of Chaoborus Licht. (Diptera: Culicidae) at Churchill, Manitoba. Mosquito News 18(3): 242-248. (11 refs., 5 figs.)

644. Jameson, A.P. 1923. Memoranda on Insecta, pp. 424-428 in W. Scoresby, Jr., Journal of a voyage to the northern whale-fishery, including researches and discoveries on the eastern coast of West Greenland made in the summer of 1822, in the Ship Baffin of Liverpool. Constable, Edinburgh. [See Scoresby 1823]

645. Jellison, W.L. 1945. Siphonaptera: the genus Oropsylla in North America. J. Parasit. 31(2): 83-97. (19 refs., 3 pls., 1 tab.)

646. Jellison, W.L. and G.M. Kohls. Siphonaptera: a list of Alaskan fleas. Publ. Hlth Rep. 54(45): 2020-2023. (3 refs.)

647. Jenkins, D.W. 1958. Ecology of arctic and subarctic mosquitoes. Proc. 10th int. Congr. Ent. (Montreal 1956). 1: 627-634. (2 refs., 18 figs., 1 tab.)

648. Jenkins, D.W. and A.S. West. 1954. Mermithid Nematode parasites in mosquitoes. Mosquito News 14(3): 138-143. (19 refs., 1 tab.)

649. Jensen, A.S. 1928a. [The fauna of Greenland...]. Festskrift, udg. af Kobenhavns Univ. i Anledn. af H.M. Kongens Fødselsdag, Kobenhavn. 88 pp. (In Danish) (90 refs., text fig.) [for English version, see Jensen 1928b]

650. Jensen, A.S. 1928b. The fauna of Greenland. Vol. I, pp. 319-355 in M. Vahl (Ed.), Greenland (3 vols.). Reitzel, Copenhagen; Oxford Univ. Press, London. (21 refs., 15 figs.)

651. Jewett, S.G., Jr. 1963. Three stoneflies (Plecoptera) from Cape Thompson, Alaska. Bull. Brooklyn ent. Soc. 58(4): 123-125. (3 refs., 1 fig.)

652. Johansen, F. 1910. General remarks on the life of insects and arachnids in northeast Greenland. The insects of the 'Danmark' expedition. Meddr Grønland 43(3): 35-54. (0 ref., 5 figs.)

653. Johansen, F. 1921. Part K: Insect life on the western Arctic coast of America. Rep. Can. Arctic Expedition 1913-1918. vol. III: Insects. 61 pp. (text refs., 1 fig., 10 pls.)

654. Johansen, F. and J.C. Nielsen. 1910. The insects of the 'Danmark' expedition. Meddr Grønland 43: 33-68. [see Johansen 1910; Nielsen 1910]

655. Johnston, E.C. 1950. Lepidoptera of the Pribilof Islands, Alaska. Lepid. News 4(3): 27-30. (2 refs.)

656. Jordan, D.S. 1899. The fur seals and fur-seal islands of the north pacific ocean. Govt. Printing office, Washington. 3 vols. [see Schwarz 1899]

657. Jordan, K. 1932. Die aus der arktischen Zone bekannten Flöhe. Fauna arct. 6: 115-118. (text refs.)

658. Jørgensen, M. 1934a. A quantitative investigation of the microfauna

communities of the soil in East Greenland. (preliminary report) Meddr Grønland 100(9). 39 pp. (34 refs., 16 tabs.)

659. Jørgensen, M. 1934b. Ein revidiertes Verzeichnis über grönländische Milben. Zool. Anz. 107(1-2): 40-47. (15 refs.)

660. Joyner, L.J.W. 1896. Note on the insects of Sukkertoppen. pp. 194-195 in H.C. Walsh et al., The last cruise of the Miranda. The Transatlantic Publishing Company. 232 pp. (0 ref.)

661. Judd, W.W. 1967. Insects from McConnell River, N.W.T. Ent. News 78(2): 50-55. (13 refs.)

662. Kalff, J. and J.E. Hobbie. 1973. An exploratory study of two arctic springs. Final report, Arct. Inst. N. Am.,project ONR-385. 24 pp. (12 refs., 2 tabs., appendix)

663. Karppinen, E., D.A. Krivolutsky, M. Koponen, L.S. Kozlovskaja, L.M. Laskova, and M. Viitasaari. 1979. List of subfossil oribatid mites (Acarina, Oribatei) of northern Europe and Greenland. Ann. ent. fenn. 45(4): 103-108. (10 refs., 16 figs., 1 tab.)

664. Kavanaugh, D.H. 1978. The nearctic species of Nebria Latreille (Coleoptera: Carabidae: Nebriini): Classification, phylogeny, zoogeography, and natural history. Unpub. Ph.D. Thesis, Univ. Alta. 2 vols. (1-521; 522-1041 pp.)

665. Kavanaugh, D.H. 1979. Studies on the Nebriini (Coleoptera: Carabidae), III. New nearctic Nebria species and subspecies, nomenclatural notes, and lectotype designations. Proc. Calif. Acad. Sci. 42(4): 87-133. (94 refs., 70 figs.)

666. Keegan, H.L. 1951. The mites of the subfamily Haemogamasinae (Acari: Laelaptidae). Proc. U.S. natn. Mus. 101(3275): 293-260. (27 refs., 55 figs., keys)

667. Keegan, H.L. and R.A. Hedeen. 1954. Collections of ectoparasitic mites from Alaska. J. Parasit. 38(4): 360-361. (5 refs.)

668. Kellogg, V.L. and S.I. Kuwana. 1900. Mallophaga from Alaskan birds. Proc. Acad. nat. Sci. Philad. 52: 151-159, 1 pl. (text refs., 5 figs.)

669. Kelsall, J.P. 1975. Warble fly distribution among some Canadian caribou. Biol. Pap. Univ. Alaska, Spec. Rep. 1: 509-517. [Proc. 1st int. Reindeer and Caribou Symp., Fairbanks 1972] (8 refs., 2 figs., 1 tab.)

670. Kelton, L.A. 1961. Synopsis of the nearctic species of Stenodema Laporte, and description of a new species from Western Canada (Hemiptera: Miridae). Can. Ent. 93(6): 450-455. (21 refs., key)

671. Kelton, L.A. 1965. Chlamydatus Curtis in North America (Hemiptera: Miridae). Can. Ent. 97(11): 1132-1144. (24 refs., 19 figs., key)

672. Kelton, L.A. 1966. Review of the species of Teratocoris Fieber, with description of a new species from the nearctic region. Can. Ent. 98(12): 1265-1271. (14 refs., 7 figs., key)

673. Kelton, L.A. 1971. Revision of the species of Trigonotylus in North America (Heteroptera: Miridae). Can. Ent. 103(5): 685-705. (30

refs., 19 figs., key)

674. Kelton, L.A. 1978. The Anthocoridae of Canada and Alaska (Heteroptera: Anthocoridae). The Insects and Arachnids of Canada, Part 4. Agric. Can. Publ. 1639. 101 pp. (68 refs., 22 maps, 123 figs., keys)

675. Kevan, P.G. 1970. High arctic insect-flower relations; the interrelationships of Arthropoda and flowers at Lake Hazen, Ellesmere Island, N.W.T., Canada. Unpub. Ph.D. thesis, Univ. Alta. 399 pp.

676. Kevan, P.G. 1972a. Floral colours in the high arctic with reference to insect-flower relations and pollination. Can. J. Bot. 50 (11): 2289-2316. (73 refs., 33 figs., 6 tabs.)

677. Kevan, P.G. 1972b. Insect pollination of high arctic flowers. J. Ecol. 60(3): 831-847. (92 refs., 1 fig., 4 tabs.)

678. Kevan, P.G. 1972c. Collembola on flowers on Banks Island, N.W.T. Quaest. ent. 8(2): 121. (1 ref.)

679. Kevan, P.G. 1972d. Heliotropism in some arctic flowers. Can. Fld Nat. 86(1): 41-44. (15 refs., 3 figs., 1 tab.)

680. Kevan, P.G. 1973a. Flowers, insects, and pollination ecology in the Canadian high arctic. Polar Rec. 16(104): 667-674. (18 refs.)

681. Kevan, P.G. 1973b. Parasitoid wasps as flower visitors in the Canadian high arctic. Sonderdruck aus Anz. für Schädlings. Pflanz. Umweltschutz 46: 3-7. (31 refs., 1 tab.)

682. Kevan, P.G. 1975. Sun tracking solar furnaces in high arctic flowers: significance for pollination and insects. Science 189(4204): 723-726. (18 refs., 2 figs., 2 tabs.)

683. Kevan, P.G. 1978a. Floral coloration, its colorimetric analysis and significance in anthecology. pp. 51-78 in A.J. Richards (Ed.), The pollination of flowers by insects. (Linnean Soc. Symp. Ser. 6.) (154 refs., 11 figs., 4 tabs., 5 pls.)

684. Kevan, P. 1978b. Anthophilous springtails (Collembola) from the Alaskan north slope, from Signy Island, Antarctica, and from near Ottawa, Ontario. Rev. Ecol. Biol. Sol. 15(3): 373-378. (19 refs.)

685. Kevan, P.G. 1979. The spectral efficiency of phototaxis for some high arctic Diptera. Arct. Alp. Res. 11(3): 349-352. (20 refs., 2 figs., 2 tabs.)

686. Kevan, P.G. and D.K. McE. Kevan. 1970. Collembola as pollen feeders and flower visitors with observations from the high arctic. Quaest.ent. 6(4): 311-326. (79 refs., 1 fig., 2 tabs.)

687. Kevan, P.G. and J.D. Shorthouse. 1970. Behavioural thermoregulation by high arctic butterflies. Arctic 23(4): 268-279. (31 refs., 3 figs., 2 tabs.)

688. Kiaer, H. 1902. Die arktischen Tenthrediniden. Fauna arct. 2: 401-438. (23 refs., 1 fig., 3 tabs.)

689. Kieffer, J.J. 1926. Chironomiden der 2. Fram-Expedition (1898-1902). Norsk ent. Tidsskr. 2: 78-89. (0 ref., 11 figs., key)

690. Kim, K.C. 1972. Louse populations of the northern fur seal (Callorhinus ursinus). Am. J. Vet. Res. 33(10): 2027-2036. (12 refs., 3 figs., 5 tabs.)

691. Kim, K.C., R.C. Chu, and G.P. Barron. 1974. Mercury in tissues and lice of northern fur seals. Bull. Environ. Contam. Toxicol. 11(3): 281-284. (19 refs., 1 tab.)

692. Kimmins, D.E. and D.G. Denning. 1951. The McLachlan types of North American Trichoptera in the British Museum. Ann. ent. Soc. Am. 44(1): 111-140. (22 refs., 25 figs.)

693. Kincaid, T. 1899. The Psychodidae of the Pacific Coast. Ent. News 10(2): 30-37. (text refs., 13 figs., key)

694. Kinosita, S. (Ed.). 1976. [Joint studies on physical and biological environments in the permafrost of Alaska and northern Canada, 1974.] Inst. Low Temp. Sci., Hokkaido Univ., Sapporo [see Tanno 1976]

695. Kirby, W. 1824. Land invertebrate animals. pp. ccxiv-ccxix in W.E. Parry, Supplement to the Appendix of Captain Parry's voyage for the discovery of a northwest passage, in the years 1819-1820, containing an account of the subjects of natural history. Appendix X. Natural history. pp. clxxx-cccx. Murray,London.(text refs.)

696. Kirby, W. 1837. Insecta. in J. Richardson, Fauna Boreali-Americana: containing descriptions of the objects of natural history collected on the late northern land expeditions, under command of Captain Sir John Franklin, R.N. Vol. 4, xxxix + 327 pp. Murray, London. [Reprinted 1978 by Arno Press, N.Y.] [see also Bethune 1870-1881, Horn 1876] (246 refs., 8 pls., 1 tab.)

697. Kissinger, D.G. 1973. A new weevil genus from America north of the Arctic circle and notes of fossils from Pliocene and Pleistocene sediments (Coleoptera: Curculionidae). Coleopts Bull. 27(4): 193-200. (10 refs., 8 figs.)

698. Klimaszewski, J. 1979. A revision of the Gymnusini and Deinopsini of the world (Coleoptera: Staphylinidae, Aleocharinae). Agric. Can. Monogr. 25. 169 pp. (88 refs., 251 figs., 22 maps, 4 tabs., keys)

699. Klöcker, A. 1903. Ein kleiner Beitrag zu dem Verzeichnisse der Lepidopteren Grönlands. Ent. Meddr 2[second ser.](2): 89-90. (text refs.)

700. Klots, A.B. 1935. Ovoviviparity in Colias? (Lepid: Pieridae). Ent. News 46(2): 58. (text refs.)

701. Klots, A.B. 1939. Brenthis aphirape (Huebner) in North America, with a new record of the species from Maine (Lepidoptera, Nymphalidae). Bull. Brooklyn ent. Soc. 34(5): 259-264. (0 ref.)

702. Klots, A.B. 1951. Holarctic butterfly speciation and subspeciation, especially in North America. Lepid. News 5(3-5): 24-27. (1 fig.)

703. Klots, A.B. 1956. Studies of New World arctic and alpine Lepidoptera. Proc. 14th int. Congr. Zool.(Copenhagen 1953): 469-470. (0 ref.)

704. Klots, A.B. 1958a. Ecological studies at Churchill, Manitoba. Proc. 10th int. Congr. Ent. (Montreal 1956) 1: 696. [abstract only]

705. Klots, A.B. 1958b. Southern extensions of arctic and subarctic insects in bogs and alpine areas. Proc. 10th int. Congr. Ent. (Montreal 1956) 1: 711 [abstract only]

706. Klots, A.B. 1961. Zoogeography in the systematics of north American Crambinae (Lepidoptera, Pyralididae). Proc. 11th int. Congr. Ent. (Vienna 1960) 1: 525-527. (1 ref.)

707. Knaben, N. 1937. Makrolepidopteren aus Nordostgrönland. (Zool. results Norwegian Sci. Exp. to East Greenland VI). Skr. Svalbard Ishavet 71: 1-30, pls. I-II. (64 refs., 10 figs., 1 map, 2 pls.)

708. Knaben, N. and E. Barca. 1937. Zoological results of the Norwegian scientific expeditions to East-Greenland. VI. 1. Makrolepidopteren aus Nordostgrönland. 2. Mikrolepidopteren aus Nordostgrönland. Skr. Svalbard Ishavet 71. 37 pp. [see Knaben 1937; Barca 1937]

709. Knight, K.L. 1948. A taxonomic treatment of the mosquitoes of Umiat, Alaska. U.S. Naval Med. Res. Inst. Proj. NM 005017, Rep. 2; 12 pp. [not seen]

710. Knight, K.L. 1951. The Aedes (Ochlerotatus) punctor sub-group in North America (Diptera, Culicidae). Ann. ent. Soc. Am. 44(1): 87-99. (5 + text refs., 8 figs., key)

711. Knight, K.L. and L.A. Jachowski. 1956. Arctic and subarctic insect pests and their control. U.S. Navy Series of Operational Briefs OPNAV P03-29. 13 pp. (14 refs.)

712. Knutson, L.V. 1973. Taxonomic revision of the aphid-killing flies of the genus Sphaerophoria in the Western Hemisphere (Syrphidae). Misc. Publs ent. Soc. Am. 9(1). 50 pp. (91 refs., 108 figs., 10 maps, key)

713. Koch, L. 1874. Arachniden. pp. 400-403 in Geographische Gesellschaft in Bremen. Die zweite deutsche Nordpolarfahrt in den Jahren 1869 und 1870. Bd. 2. (text refs., 1 pl.)

714. Kontkanen, P. 1959. Uber einige Chrysolina-Arten der sectio Pleurosticha Motsch. sensu Jacobson 1910 (Col., Chrysomelidae). Suom. hyönt. Aikak. 25(1): 27-35. (15 refs., 8 figs.)

715. Koponen, S. 1978. Notes on herbivorous insects of the birch in southern Greenland. Rep. Kevo Subarctic Res. Stn 14: 13-17. [also as Ann. Univ. Turku.] (22 refs., 1 fig., 2 tabs.)

716. Kozhanschikov, I.W. 1934. Die verbreitung der arktischen Agrotinen und die genetische Untersuchung dieser Fauna. Zool. Anz. 108 (5/6): 113-118. (4 refs., 1 fig.)

717. Kramer, P. 1897. Grönländische Milben. Zoologische Ergebnisse der von der Gesellschaft fur Erdkunde zu Berlin unter Leitung Dr. von Drygalski aus gesandten Grönlandexpedition nach Dr. Vanhoffen's Sammlung bearbeitet, VI. Bibliotheca Zoologica 20: 77-83. (text refs., 3 figs.)

718. Kramer, P. and C.J. Neumann. 1883. Acariden, wahrend der Vega-

Expedition eingesammelt. pp. 521-529, pls. 39-44 in N.E.A. Nordenskiöld, Vega-expeditionens vetenskapliga iakttagelser. (Vol. 3). (0 ref., 6 pls.)

719. Kristensen, R.M. and K. Vestergaard. 1975. [Diurnal activity of the springtail Sminthurides malmgreni Tullberg (Collembola) under Arctic summer conditions.] Ent. Meddr 43(1): 21-32. (In Danish) (13 refs., 12 figs.)

720. Krombein, K.V., P.D. Hurd, Jr., D.R. Smith, B.D. Burks, and others. 1979. Catalog of Hymenoptera in America North of Mexico. Vol. I, Symphyta and Apocrita (Parasitica), pp. 1-1198; Vol. II, Apocrita (Aculeata), pp. 1198-2210; Vol. III, Index, pp. 2211-2735. Smithsonian Institution, Washington. (text refs., 2 tabs.)

721. Kronestedt, T. 1975. Studies on species of Holarctic Pardosa groups (Araneae, Lycosidae). 1. Redescription of Pardosa albomaculata Emerton and description of two new species from North America, with comments on some taxonomic characters. Zool. Scripta 4 (5-6): 217-228. (42 refs., 9 figs., text tabs.)

722. Kryger, J.P. 1942. Tetracyclos boreios n.g. et sp. (Hym. Chalcididae, Aphelinini). Ent. Meddr 22(4): 260-263. (text refs., 3 figs., key)

723. Kumlien, L. et al. 1879. Contributions to the natural history of arctic America. Bull. U.S. natn. Mus. 15. 179 pp. [see Edwards 1879, Scudder et al. 1879.]

724. Kurentsov, A.I. 1963. Entomological fauna of Kamchatka and the Bering Arc problem in zoogeography. pp. 113-119 in J.L. Gressitt (Ed.), Pacific Basin biogeography. Bishop Museum Press, Honolulu. (Symp. 10th Pacif. Sci. Congr. 1961). 563 pp. (39 refs.)

725. Kurstak, E. (Ed.) 1979. Arctic and tropical arboviruses. (2nd int. Symp. Arctic arboviruses 1977) Academic Press, N.Y. 327 pp. [see McLean 1979]

726. Lack, D. 1934. Some insects from the Scoresby Sound region, East Greenland, with an account of the fauna of a nunatak. Ann. Mag. nat. Hist. 10(14): 599-606. (7 refs.)

727. Laird, M. 1961. A lack of avian and mammalian haematozoa in the antarctic and Canadian arctic. Can. J. Zool. 39(2): 209-213. (10 refs., 1 tab.)

728. Lange, W.H., Jr. 1940. An apparently new Alaskan Platyptilia (Lepidoptera, Pterophoridae). Can. Ent. 72(4): 84-85. (1 ref., 2 figs.)

729. Langer, T.W. 1967. [Greenland's butterfly fauna and its early investigation.] Grønland 1968 (12): 385-398. (In Danish)

730. Langer, T.W. 1968. [Investigation of the butterflies and moths of Greenland 1780-1890, an historical summary.] Naturens Verden 1968 (Feb.):56-64; (April):112-118; (June):161-176. (text + 19 refs., pls., tabs.) (In Danish)

731. Lansbury, I. 1955. Distributional records of North American Corixidae (Hemiptera: Heteroptera). Can. Ent. 87(11): 474-481. (text refs.)

732. Larson, D.J. 1975. The predaceous water beetles (Coleoptera) of Alberta: Systematics, natural history and distribution. Quaest. ent. 11(4): 245-498. (287 refs., 469 figs., 4 tabs., keys)

733. Larsson, S.G. 1951. [Spreading of insects with human help.] Naturens Verden 35(3-5): 80-94. (In Danish) (0 ref., 4 figs., 1 map)

734. Leech, H.B. 1938. Descriptions of three new species of Agabus from Hudson Bay (Coleoptera, Dytiscidae). Can. Ent. 70(6): 123-127. (2 refs., 4 figs.)

735. Leech, H.B. 1939. On some nearctic species of Agabus, with the description of a new species (Coleoptera, Dytiscidae). Can. Ent. 71(10):217-221. (16 refs., 1 fig.)

736. Leech, R.E. 1966. The spiders (Araneida) of Hazen Camp 81°49'N, 71°18'W. Quaest. ent. 2(2): 153-212. (125 refs., 65 figs., 1 tab.)

737. Leech, R.E. and J.V. Matthews, Jr. 1971. Xysticus archaeopalpus (Arachnida: Thomisidae), a new species of crab spider from Pliocene sediments in western Alaska. Can. Ent. 103(9): 1337-1340. (8 refs., 5 figs.)

738. Leech, R.E. and J.K. Ryan. 1972. Notes on Canadian arctic spiders (Araneida), mainly from Devon Island, N.W.T. Can. Ent. 104(11): 1787-1791. (9 refs., 9 figs.)

739. Lehmkuhl, D.M. 1973. A new species of Baetis (Ephemeroptera) from ponds in the Canadian Arctic, with biological notes. Can. Ent. 105(2): 343-346. (9 refs., 8 figs., 1 tab.)

740. Lehmkuhl, D.M. and C.D. Kerst. 1979. Zoogeographical affinities and identification of central arctic caddisflies (Trichoptera). Musk-Ox 25: 12-28. (28 refs., 25 figs., 3 tabs.)

741. Leng, C.W. 1919. Coccinellidae, Elateridae, Chrysomelidae and Rhynchophora (excluding Ipidae). pp. 17-18, 19-21 in Part E. Coleoptera. The Coleoptera collected by the Canadian Arctic Expedition 1913-1918. Rep. Can. Arctic Expedition 1913-1918. Vol. III: Insects. 27 pp. (0 ref.)

742. Lenz, H.W.C. 1897. Grönländische Spinnen. Zoologische Ergebnisse der Gesellschaft fur Erdkunde zu Berlin unter Leitung Dr. von Drygalski aus gesandten Grönlandexpedition nach Dr. Vanhoffen's Sammlung bearbeitet, V. Bibliotheca Zoologica 20: 73-76. (text refs., 9 figs.)

743. Letteval, U. 1962. On the Hydracarina of Greenland with a description of Lebertia (Pseudolebertia) groenlandica n. sp. Meddr Grønland 170(1). 40 pp. (48 refs., 1 fig.)

744. Leussler, R.A. and [in part] O. Bryant. 1935. Notes on the diurnal Lepidoptera of the Canadian Arctic collected by O. Bryant in the summers of 1929 to 1932. Bull. Brooklyn ent. Soc. 30(1): 1-10; (2): 42-62; (3): 115-118. (text + 2 refs.)

745. Levi, H.W. 1957. The spider genera Enoplognatha, Theridion, and Paidisca in America north of Mexico (Araneae, Theridiidae). Bull. Am. Mus. nat. Hist. 112(1): 1-124. (text refs., 421 figs., 41 maps, keys, appendix)

746. Levi, H.W. 1971. The diadematus group of the orb-weaver genus Araneus north of Mexico (Araneae: Araneidae). Bull. Mus. comp. Zool. Harv. 141(4): 131-179. (20 refs., 232 figs., 4 pls., 8 maps, key)

747. Levi, H.W. 1972. The orb-weaver genera Singa and Hypsosinga in America (Araneae: Araneidae). Psyche, Camb. 78[1971] (4): 229-256. (text refs., 106 figs., 4 maps, keys)

748. Levi, H.W. 1974. The orb-weaver genera Araniella and Nuctenea (Araneae: Araneidae). Bull. Mus. comp. Zool. Harv. 146(6): 291-316. (16 refs., 129 figs., 4 maps, 2 pls., keys)

749. Levi, H.W. 1975. Additional notes on the orb-weaver genera Araneus, Hypsosinga and Singa north of Mexico (Araneae, Araneidae). Psyche, Camb. 82(2): 265-274. (6 refs., 21 figs.)

750. Levi, H.W. 1977. The orb-weaver genera Metepeira, Kaira and Aculepeira in America north of Mexico (Araneae: Araneidae). Bull. Mus. comp. Zool. Harv. 148(5): 185-238. (17 refs., 231 figs., 4 maps, 7 pls., keys)

751. Lindberg, H. 1935. Hemiptera aus nordost Grönland. [Zoological results of the Norwegian Scientific Expeditions to East Greenland.IV]. Skr. Svalbard Ishavet 65: 11-17. (text refs., 1 fig.)

752. Lindeberg, B. 1968. Population differences in Tanytarsus gracilentus (Holmgr.) (Dipt., Chironomidae). Ann. zool. fenn. 5(1): 88-91. (14 refs., 2 figs., 1 tab.)

753. Lindegaard, C., P. Maehl, and B.H. Nielsen. 1978. Zoobenthos of lakes located within and outside the Ilimaussaq intrusion in south Greenland. Verh. int. Verein. theor. angew. Limnol. 20 (1): 159-164. (7 refs., 12 figs., 3 tabs.)

754. Lindquist, E.E. 1961. Taxonomic and biological studies of mites of the genus Arctoseius Thor from Barrow, Alaska (Acari: Aceosejidae). Hilgardia 30(11): 301-350. (25 refs., 22 pls., 13 tabs., key)

755. Lindroth, C.H. 1952. [Entomological investigations in Newfoundland.] Svensk. Naturvetensk. forskn. årsbok (Swedish National Research Council Yearbook) 1951-1952: 134-142. (In Swedish, English summary) [not seen]

756. Lindroth, C.H. 1957. The faunal connections between Europe and North America. Almquist and Wiksell, Stockholm. 344 pp. (195 + 107 + 187 refs., 61 figs., 11 diagrams, 9 tabs.)

757. Lindroth, C.H. 1958. The original faunal connections between Europe and North America. Proc. 10th int. Congr. Ent. (Montreal 1956) 1: 713. [abstract only]

758. Lindroth, C.H. 1961a. The ground beetles (Carabidae, excl. Cicindelinae) of Canada and Alaska. Part 2. Opusc. ent. Suppl. 20. pp. 1-200. (0 ref., 101 figs., keys)

759. Lindroth, C.H. 1961b. Verbindungen und barrieren in der zirkumpolaren Verbreitung der insekten. Proc. 11th int. Congr. Ent. (Vienna 1960) 1: 438-445. (6 refs., 5 figs.) [partially covered in English by Lindroth 1968b]

760. Lindroth, C.H. 1963a. The ground beetles of Canada and Alaska. Part 3. Opusc. ent. Suppl. 24. pp. 201-408. (0 ref., figs. 102-107, keys)

761. Lindroth, C.H. 1963b. The problem of late land connections in the north Atlantic area. pp. 73-85 in A. Löve and D. Löve (Eds.), North Atlantic biota and their history. Pergamon Press, Oxford. 430 pp. (28 refs., 6 figs., 2 tabs.)

762. Lindroth, C.H. 1963c. The Aleutian islands as a route for dispersal across the north Pacific. pp. 121-131 in J.L. Gressitt (Ed.), Pacific Basin biogeography. Bishop Museum Press, Honolulu. (Symp. 10th Pacif. Sci. Congr. 1961). 563 pp. (12 refs., 6 figs., 1 tab.)

763. Lindroth, C.H. 1965. Two new species of subg. Chrysobracteon (genus Bembidion) from Alaska and Siberia (Col. Carabidae). Opusc. ent. 30(1-2): 125-128. (2 refs., 2 figs.)

764. Lindroth, C.H. 1966. The ground beetles of Canada and Alaska. Part 4. Opusc. ent. Suppl. 29. pp. 409-648. (0 ref., figs. 208-328, keys)

765. Lindroth, C.H. 1968a. The ground beetles of Canada and Alaska. Part 5. Opusc. ent. Suppl. 33. pp. 649-944. (0 ref., figs. 329-480, keys)

766. Lindroth, C.H. 1968b. Distribution and distributional centres of north Atlantic insects. Bull. ent. Soc. Am. 14(2): 91-95. (6 refs., 3 figs.)

767. Lindroth, C.H. 1969a. The ground beetles of Canada and Alaska. Part 6. Opusc. ent. Suppl.34. pp. 945-1192. (838 refs., figs. 481-557, 1 tab., keys, index)

768. Lindroth, C.H. 1969b. The ground beetles of Canada and Alaska. Part 1. Opusc. ent. Suppl. 35. pp. I-XLVIII.

769. Linnaniemi, A. 1935. Collembolen aus Nordost-Grönland. (Zoological results of the Norwegian Scientific Expeditions to East Greenland. IV). Skr. Svalbard Ishavet 65: 10-35. (16 refs.)

770. Lohm, U. and T. Persson (Eds.) 1977. Soil organisms as components of ecosystems. (Proc. 6th int. Coll. Soil Zool.) Ecol. Bull. (Stockholm) 25. 614 pp. [see MacLean et al. 1977]

771. Longstaff, T.G. 1932. An ecological reconnaissance in west Greenland. J. anim. Ecol. 1(2): 119-142, pl. VI-IX. (16 refs., 4 pls., 6 tabs.)

772. Longstaff, T.G. 1936. Entomological note on Mr. J.M. Wordie's expedition to N.W. Greenland and E. Baffinland in 1934. Ann. Mag. nat. Hist. 10(18): 531-533. (0 ref., 1 tab.)

773. Løppenthin, B. 1943. Systematic and biologic notes on the long-tailed skua Stercorarius longicaudus Vieill. Meddr Grønland 131(12). 26 pp. (48 refs., 4 figs., 2 tabs.)

774. Lundbeck, W. 1891. [Entomological investigations in west Greenland 1889 and 1890.] Meddr Grønland 7(4): 107-144, pls. V-VII. (In Danish) (0 ref., 18 figs.)

775. Lundbeck, W. 1891. [Notes on the insect fauna of Greenland.] Ent.

Meddr 3[original series]: 45-52. (In Danish) (0 ref.)

776. Lundbeck, W. 1896a. Coleoptera Groenlandica. Vidensk. Meddr dansk naturh. Foren. 196-219. (In Danish) (text refs.)

777. Lundbeck, W. 1896b. Hymenoptera groenlandica. Vidensk. Meddr dansk naturh. Foren. 220-251. (In Danish) (text refs.)

778. Lundbeck, W. 1896c. [Insects of East Greenland. Catalogue of insects collected.] Meddr Grønland 19: 105-120. (text refs.)

779. Lundbeck, W. 1898. Diptera groenlandica. Vidensk. Meddr dansk naturh. Foren. 5(10): 236-314. (In Danish) (text refs., 20 figs.)

780. Lundbeck, W. 1899-1928. Conspectus fauna groenlandicae. Reitzel, Copenhagen (Meddr Grønland 21-23). [see Henriksen and Lundbeck 1917]

781. Lundbeck, W. 1901. Diptera groenlandica. Vidensk. Meddr dansk naturh. Foren. 6(2): 281-316. (In Danish) (text refs., 4 figs.)

782. Lutz, F.E. 1916. The geographic distribution of Bombidae (Hymenoptera), with notes on certain species of boreal America. Bull. Am. Mus. nat. Hist. 35(26): 501-521. (0 ref., 2 figs.)

783. Lyman, H.H. 1884. List of Lepidoptera collected by Dr. Robert Bell in Hudson's Strait in 1884. Appendix 5, p. 61DD in R. Bell, Observations on the geology, minerology, zoology and botany of the Labrador Coast, Hudson's Strait and Bay. Geol. Surv. Can. A. Rep.1882-1883-1884. Pt. DD. (0 ref.)

784. Lyons, E.T., M.C. Keyes, and J. Conlogue. 1978. Activities of Dichlorvos or Disophenol against the hookworm (Uncinaria lucasi) and sucking lice of northern fur seal pups (Callorhinus ursinus) on St. Paul Island, Alaska. J. Wildl. Dis. 14(4): 455-464. (11 refs., 4 tabs.)

785. MacDonald, S.D. 1953. Report on biological investigations at Alert, N.W.T. Bull. natn. Mus. Can. 128: 241-256. (0 ref., 1 fig., text tabs.)

786. MacDonald, S.D. and D.F. Parmelee. 1962. Feeding behaviour of the Turnstone in arctic Canada. Br. Birds 55(7): 241-244, 1 pl. (0 ref., 1 pl.)

787. MacGillivray, A.D. 1891. A catalogue of the Thysanoura of North America. Can. Ent. 23(12): 267-276. (28 refs.)

788. MacGillivray, A.D. 1919. Part G: Hymenoptera and plant galls. The saw-flies (Tenthredinoidea) collected by the Canadian Arctic Expedition, 1913-1918. Rep. Can. Arctic Expedition 1913-1918. Vol. III: Insects. pp. 3-19. (0 ref.)

789. MacGillivray, A.D. 1921. New saw-flies from the Pribilof Islands, Alaska. pp. 188-192 in G.D. Hanna et al., Insects of the Pribilof Islands, Alaska. Proc. Calif. Acad. Sci. ser. 4, 11 (14): 153-195. (0 ref.)

790. MacInnes, K.L. 1972. Reproductive ecology of five arctic species of Pedicularis (Scrophulariaceae). Unpub. Ph.D. thesis, Univ. Western Ontario, London, Ont. 235 pp.

791. MacKay, M.R. 1972. The larvae of Canadian arctic Noctuidae (Lepidoptera). Can. Ent. 104(6): 859-872. (3 refs., 7 figs., key)

792. Mackay, M.R. and J.A. Downes. 1969. Arctic Olethreutinae (Lepidoptera, Tortricidae). Can. Ent. 101(10): 1048-1053. (2 refs., 2 figs.)

793. MacLean, S.F.,Jr. 1971a. Population ecology and energetics of tundra soil arthropods. pp. 126-130 in J. Brown and S. Bowen (Eds.), The structure and function of the tundra ecosystem, Vol. I. U.S. Tundra Biome Program, U.S. International Biological Program. (0 ref., 4 tabs.)

794. MacLean, S.F., Jr. 1971b. Temporal patterns in the emergence of tundra crane flies. Proc. 22nd Alaska Sci. Conf.: 94.[abstract only]

795. MacLean, S.F., Jr. 1972. Population dynamics and energetics of tundra crane flies (Diptera: Tipulidae). p. 144 in Proc. 1972 Tundra Biome Symposium. [abstract only]

796. MacLean, S.F., Jr. 1973. Life cycle and growth energetics of the arctic crane fly Pedicia hannai antenatta. Oikos 24(3): 436-443. (18 refs., 6 figs., 3 tabs.)

797. MacLean, S.F., Jr. 1974. Primary production, decomposition and the activity of soil invertebrates in tundra ecosystems: A hypothesis. pp. 197-206 in A.J. Holding, O.W. Heal, S.F. MacLean, Jr., and P.W. Flanagan (Eds.), Soil organisms and decomposition in tundra. Tundra Biome Steering Committee, Stockholm, Sweden. 398 pp. (6 refs., 7 figs., 1 tab.)

798. MacLean, S.F., Jr. 1975a. Ecology of tundra invertebrates at Prudhoe Bay, Alaska. pp. 115-123 in J. Brown (Ed.), Ecological investigations of the tundra biome in the Prudhoe Bay region, Alaska. Univ. Alaska Biol. pap. spec. rep. 2. 215 pp. (7 refs.)

799. MacLean, S.F., Jr. 1975b. Ecological adaptations of tundra invertebrates. pp. 269-300 in J. Vernberg (Ed.), Proc. AIBS Symposium on physiological adaptation to the environment. Amhurst, Mass. (96 refs., 4 figs., 2 tabs.)

800. MacLean, S.F., Jr., V. Behan, and A. Fjellberg. 1978. Soil Acari and Collembola from Chaun Bay, Northern Chukotka, USSR. Arct. Alp. Res. 10(3): 559-568. (22 refs., 2 figs., 4 tabs.)

801. MacLean, S.F., G.K. Douce, E.A. Morgan, and M.A. Skeel. 1977. Community organisation in the soil invertebrates of Alaskan arctic tundra. pp. 90-101 in U. Lohm and T. Persson (Eds.), Soil organisms as components of ecosystems. Ecol. Bull. (Stockholm) 25. 614 pp. (21 refs., 6 figs., 2 tabs.)

802. MacLean, S.F., Jr. and F.A. Pitelka. 1971. Seasonal patterns of abundance of tundra arthropods near Barrow. Arctic 24(1): 19-40. (22 refs., 8 figs., 4 tabs.)

803. Madsen, B.L. 1959. [Limnological investigations in southwest Greenland.] Flora fauna 65(2): 41-48. (In Danish, English summary) (6 refs., 8 figs.)

804. Madsen, H. 1936. Investigations on the shore fauna of East Greenland with a survey of the shores of other arctic regions. Meddr

Grønland 100(8). 79 pp. (161 refs., 17 figs., 2 tabs.)

805. Maher, W.J. 1970. Ecology of the long-tailed jaeger at Lake Hazen, Ellesmere Island. Arctic 23(2): 112-129. (23 refs., 3 figs., 4 tabs.)

806. Malaise, R.E. 1933. Eine neue Blattwespe aus Ost-Grönland. (Zoological results of the Norwegian Scientific Expeditions to East Greenland. II). Skr. Svalbard Ishavet 53: 3-4. (0 ref., 1 fig.)

807. Malloch, J.R. 1913. A revision of the species of Agromyza Fallen, and Cerodontha Rondani (Diptera). Ann. ent. Soc. Am. 6(3): 269-336, pls. XXVIII-XXXI. (text refs., 39 figs., key)

808. Malloch, J.R. 1919. Part C: Diptera (excluding the Tipulidae and Culicidae). The Diptera collected by the Canadian Expedition, 1913-1918. Rep. Can. Arctic Expedition 1913-1918. Vol. III: Insects. pp. 34-90. (0 ref., 38 figs., keys)

809. Malloch, J.R. 1920a. Descriptions of new North American Anthomyiidae (Diptera). Trans. Am. ent. Soc. 46(2): 133-196, pls. IV-VI. (text refs., 50 figs.)

810. Malloch, J.R. 1920b. A synoptic revision of the anthomyoid genus Hydrophoria Robineau-Desvoidy (Diptera). Can. Ent. 52(11): 253-257. (0 ref., key)

811. Malloch, J.R. 1921a. A synopsis of the North American species of the genus Helina R.-D., sens. lat. (Diptera, Anthomyiidae). Can. Ent. 53(5): 103-109. (0 ref., key)

812. Malloch, J.R. 1921b. Dipterous insects of the family Anthomyiidae from the Pribilof Islands, Alaska. pp. 178-182 in G.D. Hanna et al., Insects of the Pribilof Islands, Alaska. Proc. Calif. Acad. Sci. ser. 4, 11(14): 153-195. (0 ref., key)

813. Malloch, J.R. 1923. Diptera (except Tipulidae, Rhyphidae and Calliphoridae). pp. 170-227 in W.L. McAtee (Ed.), A biological survey of the Pribilof Islands, Alaska. Part II, Insects, arachnids and chilopods of the Pribilof Islands, Alaska. N. Am. Fauna 46: 129-255. (text refs., 37 figs., keys)

814. Malloch, J.R. 1934. III. Chironomidae, Sciaridae, Phoridae, Syrphidae, Piophilidae, Helomyzidae, Calliphoridae, Oestridae, and Tachinidae. The exploration of Southampton Island, Hudson Bay, by G.M. Sutton. Mem. Carnegie Mus. 12, part II, sect. 4: 13-32. (text refs., 14 figs., key)

815. Malloch, J.R. 1935. Exotic Muscaridae (Diptera). XXXVIII. Ann. Mag. nat. Hist. 10(15): 242-266. (text refs., 22 figs., key)

816. Mank, E.W. 1940. A new species of Haliplus. Psyche, Camb. 47(2-3): 57-59. (0 ref., 2 figs.)

817. Mannerheim, C.G. 1852. Zweiter Nachtrag zur Kaefer-fauna der nord-amerikanischen Laender des russischen Reiches. Byull. Mosk. Obshch. Ispyt. Prir. 25(1): 283-387.

818. Mannerheim, C.G. 1853. Dritter Nachtrag zur Kaefer-fauna der nord-amerikanischen Laender des russischen Reiches. Byull. Mosk. Obshch. Ispyt. Prir. 26(2): 95-273.

819. Mannheims, B. and B. Theowald. 1971. Die Tipuliden von Grönland (Diptera, Tipulidae). Bonn. Zool. Beitr. 22(3-4): 332-342. (40 refs., 4 figs.)

820. Martin, M. and T.W. Barry. 1978. Nesting behavior and food habits of parasitic jaegers at Anderson River delta, Northwest Territories. Can. Fld Nat. 92(1): 45-50. (11 refs., 1 fig., 2 tabs.)

821. Marx, E.J.F. 1957. A review of the subgenus *Donacia* in the Western Hemisphere (Coleoptera, Donaciidae). Bull. Am. Mus. nat. Hist. 112(3): 191-278. (57 refs., 54 figs., keys)

822. Marx, G. 1892. A contribution to the study of the spider fauna of the arctic regions. Proc. ent. Soc. Wash. 2(2): 186-201. (25 text refs.)

823. Mason, W.R.M. 1955. A revision of the nearctic Cteniscini (Hymenoptera: Ichneumonidae). 1. *Eudiaborus* Kerrich and a new genus. Can. J. Zool. 33(1): 18-73. (5 refs., 73 figs., key)

824. Mason, W.R.M. 1956. A revision of the nearctic Cteniscini (Hymenoptera: Ichneumonidae) II. *Acrotomus* Hlgr. and *Smicroplectus* Thom. Can. J. Zool. 34(2): 120-151. (text refs., 18 figs., key)

825. Mason, W.R.M. 1958. Distributional problems in Alaska. Proc. 10th int. Congr. Ent. (Montreal 1956) 1: 703-710. (11 refs., 11 figs.)

826. Mason, W.R.M. 1959. A revision of the nearctic Exenterini (Hymenoptera: Ichneumonidae) III *Exyston* Schiodte. Can. J. Zool. 37(6): 1067-1115. (4 refs., 22 figs., 2 tabs., key)

827. Mason, W.R.M. 1964a. Regional color patterns in the parasitic Hymenoptera. Can. Ent. 96(1-2): 132-134. (0 ref.)

828. Mason, W.R.M. 1964b. Southward extension of the arctic zone. Can. Ent. 96(1-2): 140. (0 ref.)

829. Masters, J.H. 1971a. A record of *Pieris rapae* (Lepidoptera: Pieridae) from Eskimo Point, N.W.T., Canada. Bull. Assoc. Minn. Ent. 4(4): 53 (text ref.)

830. Masters, J.H. 1971b. The butterflies of Churchill, Manitoba. Mid-Cont. Lepid. Ser. 2(25): 1-13. (7 refs., 1 fig., 1 tab.)

831. Masters, J.H. 1972. Ecological and distributional notes on butterflies of the genus *Erebia* in Manitoba. Mid-Cont. Lepid. Ser. 4 (52): 1-10. (18 refs., 5 figs., key)

832. Matheson, R. 1930. Distribution notes on Culicidae (mosquitoes). Bull. Brooklyn ent. Soc. 25(5): 291-294. (0 ref.)

833. Mathis, W.N. 1973. A review of the genus *Borboropsis* (Diptera: Heleomyzidae). Pan-Pacif. Ent. 49(4): 373-377. (6 refs., 1 fig., key)

834. Mathis, W.N. 1979. Studies of Ephydrinae (Diptera: Ephydridae) II: Phylogeny, classification, and zoogeography of nearctic *Lamproscatella* Hendel. Smithson. Contribs Zool. 295. 41 pp. (43 refs., 52 figs., 2 tabs., keys)

835. Matthews, J.V., Jr. 1970. Two new species of Micropeplus from the Pliocene of western Alaska with remarks on the evolution of Micropeplinae (Coleoptera: Staphylinidae). Can. J. Zool. 48(4): 779-788. (8 refs., 5 figs.)

836. Matthews, J.V., Jr. 1974a. A preliminary list of insect fossils from the Beaufort formation, Meighen Island, District of Franklin. Geol. Surv. Can. Pap. 74-1A: 203-206. (19 refs., 1 tab.)

837. Matthews, J.V., Jr. 1974b. Quaternary environments at Cape Deceit (Seward Peninsula, Alaska): evolution of a tundra ecosystem. Geol. Soc. Am. Bull. 85(9): 1353-1384. (126 refs., 17 figs., 6 tabs.)

838. Matthews, J.V., Jr. 1975. Incongruence of macrofossil and pollen evidence: A case from the late Pleistocene of the Northern Yukon coast. Geol. Surv. Can. Pap. 75-1B: 139-146. (28 refs., 3 figs., 2 tabs.)

839. Matthews, J.V., Jr. 1976a. Insect fossils from the Beaufort formation: Geological and biological significance. Geol. Surv. Can. Pap. 76-1B: 217-227. (25 refs., 2 figs., 1 tab.)

840. Matthews, J.V., Jr. 1976b. Evolution of the subgenus Cyphelophorus (genus Helophorus, Hydrophilidae, Coleoptera): description of two new fossil species and discussion of Helophorus tuberculatus Gyll. Can. J. Zool. 54(5): 652-673. (44 refs., 11 figs.)

841. Matthews, J.V., Jr. 1977a. Tertiary Coleoptera fossils from the North American arctic. Coleopts Bull. 31(4): 297-308. (37 refs. (+ 5 in erratum), 2 figs., 1 tab.)

842. Matthews, J.V., Jr. 1977b. Coleoptera fossils: their potential value for dating and correlation of late Cenozoic sediments. Can. J. Earth Sci. 14(10): 2339-2347. (33 refs., 4 figs., 1 tab.)

843. Matthews, J.V., Jr. 1979a. Tertiary and Quaternary environments: historical background for an analysis of the Canadian insect fauna. pp. 31-86 in H.V. Danks (Ed.), Canada and its insect fauna. Mem. ent. Soc. Can. 108. 573 pp. (318 refs., 9 figs.)

844. Matthews, J.V., Jr. 1979b. Beringia during the late Pleistocene: arctic steppe or discontinuous herb tundra? A review of the paleontological evidence. Geol. Surv. Can. Open-File Rep. 649. 60 pp. (157 refs., 6 figs., 1 tab.)

845. Matthews, J.V., Jr. 1979c. Fossil beetles and the late Cenozoic history of the tundra environment. pp. 371-378 in J. Gray and A.J. Boucot (Ed.), Historical biogeography, plate tectonics, and the changing environment. Oregon State Univ. Press, Corvallis. (49 refs., 3 figs.)

846. May, E. 1932. Die Odonaten des arktischen Gebietes. Fauna arct. 6: 175-182. (text refs.)

847. Mayfield, H.F. 1978. Undependable breeding conditions in the Red Phalarope. Auk 95(3): 590-592. (8 refs.)

848. McAlpine, J.F. 1961a. Variation, distribution and evolution of the Tabanus (Hybomitra) frontalis complex of horse flies (Diptera: Tabanidae). Can. Ent. 93(10): 894-924. (27 refs., 31 figs., 2 tabs.)

849. McAlpine, J.F. 1961b. Preliminary report on the insects of the Isachsen area, Ellef Ringnes Island, 1960. Arct. Circ. 14(4): 71-72. (0 ref.)

850. McAlpine, J.F. 1964. Arthropods of the bleakest barren lands: composition and distribution of the arthropod fauna of the northwestern Queen Elizabeth Islands. Can. Ent. 96(1-2): 127-129. (2 refs., 1 tab.)

851. McAlpine, J.F. 1965a. Insects and related terrestrial invertebrates of Ellef Ringnes Island. Arctic 18(2): 73-103. (100 refs., 7 figs., 2 tabs.)

852. McAlpine, J.F. 1965b. Observations on anthophilous Diptera at Lake Hazen, Ellesmere Island. Can. Fld Nat. 79(4): 247-252. (6 refs.)

853. McAlpine, J.F. 1977. A revised classification of the Piophilidae, including 'Neottiophilidae' and 'Thyreophoridae' (Diptera: Schizophora). Mem. ent. Soc. Can. 103. 66 pp. (122 refs., 2 dendrograms + 58 figs., 2 tabs., keys)

854. McAtee, W.L. (Ed.) 1923a. A biological survey of the Pribilof Islands, Alaska. Part II. Insects, arachnids and chilopods of the Pribilof Islands, Alaska. N. Am. Fauna 46: 129-255. [see Alexander, Banks, Chamberlin, Ferris, Forbes, Hebard, Malloch, McAtee, Patch, Viereck, Walton, Wickham]

855. McAtee, W.L. 1923b. Introduction, pp. 129-138 in W.L. McAtee (Ed.), A biological survey of the Pribilof Islands, Alaska. Part II, Insects, arachnids and chilopods of the Pribilof Islands, Alaska. N. Am. Fauna 46: 129-255. (text refs., 1 tab.)

856. McAtee, W.L. 1923c. Apterygota. p. 139 in W.L. McAtee (Ed.), A biological survey of the Pribilof Islands, Alaska. Part II, Insects, arachnids and chilopods of the Pribilof Islands, Alaska. N. Am. Fauna 46: 129-255. (text refs.)

857. McAtee, W.L. 1923d. Anoplura. p.142 in W.L. McAtee (Ed.), A biological survey of the Pribilof Islands, Alaska. Part II, Insects, arachnids and chilopods of the Pribilof Islands, Alaska. N. Am. Fauna 46: 129-255. (text refs.)

858. McAtee, W.L. 1923e. Heteroptera. p. 145 in W.L. McAtee (Ed.), A biological survey of the Pribilof Islands, Alaska. Part II, Insects, arachnids and chilopods of the Pribilof Islands, Alaska. N. Am. Fauna 46: 129-255. (text refs.)

859. McClure, H.E. 1937. Barren land bugs. Beaver 267(4): 16-21; 61-62. (0 ref., text figs.)

860. McClure, H.E. 1943. Aspection in the biotic communities of the Churchill area, Manitoba. Ecol. Monogr. 13(1): 1-35. (10 refs., 24 figs., 19 tabs.)

861. McClymont, D.M. 1968. Phenology of high arctic Calliphoridae. Unpub. B.Sc. thesis, Queen's Univ., Kingston, Ont. 61 pp.

862. McCook, J. 1963. Mosquitoed into eloquence. Beaver 294 (summer 1963): 26-29. (0 ref.)

863. McDuffie, W.M., H.F. Cross, C.R. Twinn, A.W.A. Brown., and C.N. Husman. 1949. The effectiveness of DDT and other insecticides

as larvicides against arctic species of Aedes. Mosquito News 9(4): 145-149. (4 refs., 3 tabs.)

864. McDuffie, W.C., J.F. Sharp, H.F. Cross, C.R. Twinn, and D.M. Wilson. 1949. The effectiveness of prehatching treatments for the control of arctic mosquitoes. Mosquito News 9(2): 51-56. (6 refs., 2 tabs.)

865. McDunnough, J.H. 1928a. Insects from Baffin Island. Bull. natn.Mus. Can. 53 (Biol. Ser. no. 15): 118-120. (0 ref.)

866. McDunnough, J.H. 1928b. Notes on Canadian diurnal Lepidoptera. Can. Ent. 60(11): 266-275. (0 ref.)

867. McDunnough, J.H. 1932. Notes on agrotid genera with descriptions of new species (Lepid.). Can. Ent. 64(5): 104-112. (text refs., 1 fig.)

868. McDunnough, J.H. 1936. A new arctic baetid (Ephemeroptera). Can. Ent. 68(2): 32-34. (0 ref., 6 figs.)

869. McDunnough, J.H. 1937. Critical notes on the Canadian species of the genus Erebia (Lepid.). Can. Ent. 69(1): 14-18. (text refs.)

870. McDunnough, J.H. 1938a. New species of North American Ephemeroptera with critical notes. Can. Ent. 70(2): 23-34. (text refs., 2 figs., 1 pl.)

871. McDunnough, J.H. 1938b. Some apparently new Eucosmidae (Lepid.). Can. Ent. 70(5): 90-100. (0 ref., [14] figs.)

872. McDunnough, J.H. 1939a. Pterophorid descriptions and notes: Lepid. Can. Ent. 71(5): 109-112. (0 ref., 4 figs.)

873. McDunnough, J.H. 1939b. New North American Geometridae with notes I. Can. Ent. 71(9): 187-192. (0 ref.)

874. McDunnough, J.H. 1939c. New North American Geometridae with notes II. Can. Ent. 71(12): 249-258. (0 ref., 1 fig.)

875. McDunnough, J.H. 1944. Revision of the North American genera and species of the Phalaenid subfamily, Plusiinae (Lepidoptera). Mem. sth Calif. Acad. Sci. 2(2): 175-232. (text refs., 6 pls., keys)

876. McDunnough, J.H. 1949. Revision of the North American species of the genus Eupithecia (Lepidoptera, Geometridae). Bull. Am. Mus. nat. Hist. 93: 533-728, pls. 26-32. (text refs., 20 figs., 7 pls.)

877. McDunnough, J.H. 1950. Species of Euxoa of eastern North America, with particular reference to genitalic characters (Lepidoptera, Phalaenidae). Bull. Am. Mus. nat. Hist. 95(6): 355-408. (text refs., 11 figs., key)

878. McGuffin, W.C. 1967a. Holarctias (Lepidoptera: Geometridae). Can. Ent. 99(7): 716-721. (5 refs., 25 figs.)

879. McGuffin, W.C. 1967b. Guide to the Geometridae of Canada (Lepidoptera). I. Subfamily Sterrhinae. Mem. ent. Soc. Can. 50. 67 pp. (59 refs., 102 figs., keys)

880. McGuffin, W.C. 1973. The Rheumaptera of North America(Lepidoptera:

Geometridae). Can. Ent. 105(3): 383-398. (29 refs., 59 figs., key)

881. McIlroy, W.D. 1934. IV. Hymenoptera. The exploration of Southampton Island, Hudson Bay, by G.M. Sutton. Mem. Carnegie Mus. 12, part II, sect. 4: 33. (text refs.)

882. McLachlan, R. 1877. The insects of the arctic expedition. Entomologist's mon. Mag. 13: 181. (0 ref.)

883. McLachlan, R. 1878. Insecta and Arachnida. Appendix VI, pp. 234-239 in G.S. Nares, Narrative of a voyage to the Polar Sea during 1875-1876 in H.M. Ships 'Alert' and 'Discovery', with notes on the Natural History edited by H.W. Feilden, F.G.S., C.M.C.S., F.R.G.S., naturalist to the Expedition. Vol. 2. (0 ref.) [summary of McLachlan 1879]

884. McLachlan, R. 1879. Report on the Insecta (including Arachnida) collected by Captain Feilden and Mr. Hart between the parallels of 78° and 83° North Latitude, during the recent Arctic Expedition. J. Linn. Soc. (Zool.) 14: 98-122. (text refs., 1 fig.)

885. McLachlan, R. 1887. Report on the Neuroptera collected by Baron Nordenskiöld during the voyage of the "Vega" in 1878-1879. pp. 81-85 in N.E.A. Nordenskiöld, Vega-expeditionens vetenskapliga iakttagelser (Vol. 4). (text refs.)

886. McLean, D.M. 1975. Mosquito borne arboviruses in arctic America. Med. Biol. 53(5): 264-270. (29 refs., 1 fig.)

887. McLean, D.M. 1979. Arbovirus vectors in the Canadian arctic. pp. 7-19 in E. Kurstak (Ed.), Arctic and tropical arboviruses. (Proc. 2nd int. Symp. arctic arboviruses, Mont Gabriel 1977). Academic Press, N.Y. 327 pp. (18 refs., 7 tabs.)

888. McLean, D.M., P.N. Grass, and B.D. Judd. 1977. California encephalitis virus transmission by arctic and domestic mosquitoes. Archives Virology 55(1-2): 39-45. (12 refs., 2 tabs.)

889. McLean, D.M., P.N. Grass, and B.D. Judd. 1978. Bunyavirus and flavivirus transmission by arctic mosquitoes. J. med. Microbiol. 11(4): 19. [abstract only]

890. McLean, D.M., P.N. Grass, B.D. Judd, L.V. Ligate, and K.K. Peter. 1977a. Bunyavirus isolations from mosquitoes in the western Canadian arctic. J. Hyg.,Camb. 79(1): 61-71. (13 refs., 1 fig., 5 tabs.)

891. McLean, D.M., P.N. Grass, B.D. Judd, K.J. Stolz, and K.K. Wong. 1978. Transmission of Northway virus and St. Louis encephalitis virus by arctic mosquitoes. Archives Virology 57(4): 315-322. (15 refs., 3 tabs.)

892. Medler, J.T. 1958. A review of the genus Scleroracus in North America (Homoptera: Cicadellidae). Ann. ent. Soc. Am. 51(3): 230-241. (11 refs., 30 figs., keys)

893. Meijere, J.C.H. de 1910. Die Dipteren der arktischen Inseln. Fauna arct. 5: 13-72. (57 refs.)

894. Meinert, F. 1897. Neuroptera, Pseudoneuroptera, Thysanopoda, Mallophaga, Collembola, Suctoria, Siphunculata Groenlandica. Vidensk. Meddr dansk naturh. Foren. 48: 154-177. (In Danish) (text refs.)

895. Meinertz, T. 1973. The occurrence of Mitopus morio (Fabr.) Phalangiidae, Opiliones in Greenland and in other northern regions. Meddr Grønland 191(7). 16 pp., pls. 1-5. (24 refs., 5 figs., 5 pls., 1 tab.)

896. Meldorf, G. 1911. [Clinical and parasitological investigations in Greenland.] Meddr Grønland 47: 1-145. (In Danish, English summary) (text refs.)

897. Mercanton, P.L. 1920. Les moustiques. (Travaux de l'escouade occidentale. Pt. III) pp. 305-307 in A. Quervain and P.L. Mercanton (Eds.), Ergebnisse der Schweizerischen Grönland-expedition, 1912-1913. Schweiz. Naturforsch. Ges. Denkschrift. 53: 229-307. (0 ref.)

898. Mercanton, P.L. 1925. Notice entomologique. Moustiques. p. 268 in A. Quervain and P.L. Mercanton, Résultats Scientifiques de l'expédition Suisse au Groenland 1912-1913. Meddr Grønland 59: 55-271. (0 ref.)

899. Merritt, R.W. and B.V. Peterson. 1976. A synopsis of the Micropezidae (Diptera) of Canada and Alaska, with descriptions of 4 new species. Can. J. Zool. 54(9): 1488-1506. (20 refs., 35 figs., 7 maps)

900. Meyrick, E. 1934. A new micro-lepidopteron (Lepidoptera Tineina) from Greenland. Entomologist 67: 59. (0 ref.)

901. Mikkola, K. 1968. Transportation by air of Nomophila noctuella Schiff. (Lep., Pyralididae) to Greenland from North America in 1954. Ent. Meddr 36(5): 509-510. (2 ref.)

902. Miller, C.D.F. 1958. Distributional and nomenclatorial problems in some forms of Vespula in North America (Hymenoptera: Vespidae). Proc. 10th int. Congr. Ent. (Montreal 1956) 1: 257-264. (2 refs., 14 figs.)

903. Miller, C.D.F. 1961. Taxonomy and distribution of nearctic Vespula. Can. Ent. Suppl. 22. 52 pp. (text + 11 refs., 84 figs., 15 maps, key)

904. Miller, C.D.F. 1970. The nearctic species of Pnigalio and Sympiesis (Hymenoptera: Eulophidae). Mem. ent. Soc. Can. 68. 121 pp. (58 refs., 262 figs., 18 maps, 18 tabs., keys)

905. Miller, L.A. 1951. Observations on the bionomics of some northern species of Tabanidae (Diptera). Can. J. Zool. 29(3): 240-263. (23 refs., 13 figs., 2 tabs.)

906. Miller, L.D. 1961. North American annual summary for 1960. Zone VIII: The far north. News Lepid. Soc. 1961 (4): 13. (0 ref.)

907. Miller, L.D. 1965. North American annual summary for 1964. Zone VIII: The far north. News Lepid. Soc. 1965 (3): 11-12. (0 ref.)

908. Miller, L.D. 1967. Annual field season summary for 1966. Zone VIII: The far north. News Lepid. Soc. 1967 (3): 15. (0 ref.)

909. Miller, M.C. and J.E. Hobbie. 1976. RATE [Research on arctic tundra environments]. The Toolik Lake program. Arct. Bull. 2(9): 161-164. (0 ref., 2 figs.)

910. Miller, M.C., J.R. Vestal, S. Mozley, M. Butler,and J.E. Hobbie. 1977.

Effects of Prudhoe crude oil spills on coastal tundra ponds. U.S. Environ. Protect. Agency, Decision Ser., Energy/Env. II., EPA-600/9-77-012: 521-529. [not seen]

911. Milliron, H.E. 1971. A monograph of the western hemisphere bumblebees (Hymenoptera: Apidae; Bombinae). The genera Bombus and Megabombus subgenus Bombias. Mem. ent. Soc. Can. 82. 80 pp. (text refs., 14 figs., 6 maps, 11 pls., keys)

912. Milliron, H.E. 1973. A monograph of the western hemisphere bumblebees (Hymenoptera: Apidae; Bombinae). II. The genus Megabombus subgenus Megabombus. Mem. ent. Soc. Can. 89. 237 pp. (text refs., 29 maps, 4 pls., keys)

913. Milliron, H.E. and D.R. Oliver. 1966. Bumblebees from northern Ellesmere Island with observations on usurpation by Megabombus hyperboreus (Schönh.) (Hymenoptera: Apidae). Can. Ent. 98(2): 207-213. (11 refs., 6 figs.)

914. Mills, H.B. and W.R. Richards. 1953. Collembola from arctic and boreal Canada. J. Kans. ent. Soc. 26(2): 53-59. (0 ref., 11 figs.)

915. Minns, C.K. 1977. Limnology of some lakes on Truelove Lowland. pp. 569-585 in L.C. Bliss (Ed.), Truelove lowland, Devon Island, Canada: A high arctic ecosystem. Univ. Alta. Press, Edmonton. 714 pp. (23 refs., 9 figs., 6 tabs.)

916. Montgomery, B.E. 1958. Studies of the ecology of the bumble bees of the arctic slope of Alaska. Final report, Arct. Inst. N. Am. project ONR-158, ONR-180. 16 pp. (0 ref., 7 tabs.)

917. Moore, J.W. 1979. Some factors influencing the diversity and species composition of benthic invertebrate communities in twenty arctic and subarctic lakes. Int. Revue ges. Hydrobiol. 63(6): 757-771. (40 refs., 5 figs., 4 tabs.)

918. Moore, J.W. and I.A. Moore. 1974. Food and growth of Arctic Char Salvelinus alpinus (L.), in the Cumberland Sound area of Baffin Island. J. Fish. Biol. 6(1): 79-92. (30 refs., 8 figs., 4 tabs.)

919. Morihara, D.K. and W.P. McCafferty. 1979. Subspecies of the transatlantic species, Baetis macani (Ephemeroptera: Baetidae). Proc. ent. Soc. Wash. 81(1): 34-37. (6 refs., 9 figs.)

920. Morrison, H. 1925. Classification of scale insects of the subfamily Orthezinae. J. Agric. Res. 30(2): 97-154. (text refs., 42 figs., keys)

921. Morrison, H. 1952. Classification of the Ortheziidae. Tech. Bull. U.S. Dep. Agric. 1052. 80 pp. (45 refs., 27 figs., 1 pl., keys)

922. Mosely, M.E. 1928. Oxford University Greenland Expedition, 1928. - Trichoptera and Ephemeroptera of Greenland: additional records made by the Oxford University Expedition to Kugssuk, Godthaab Fjord, W. Greenland. Ann. Mag. nat. Hist. 10(4): 501-509. (13 refs., 6 figs.)

923. Mosely, M.E. 1931. Apatania arctica Bohem.♂. Entomologist 64: 34. (text refs.)

924. Mosely, M.E. 1932. Oxford University Greenland Expedition 1928: A

correction. Ann. Mag. nat. Hist. 10(9): 573-574, 1 pl. (text refs., 5 figs.)

925. Mosquin, T. and J.E.H. Martin. 1967. Observations on the pollination biology of plants on Melville Island, N.W.T., Canada. Can. Fld Nat. 81(3): 201-205. (9 refs., 1 fig., 1 tab.)

926. Mozley, S.C. 1978. Effects of experimental oil spills on Chironomidae in Alaskan tundra ponds. Verh. int. Verein. theor. angew Limnol. 20: 1941-1945. (4 refs., 4 figs., 1 tab.)

927. Mozley, S.C. and M.G. Butler. 1978. Effects of crude oil on aquatic insects of tundra ponds. Arctic 31(3): 229-241. (18 refs., 2 figs., 2 tabs.)

928. Muesebeck, C.F.W. 1921. A revision of the North American species of ichneumon-flies belonging to the genus Apanteles. Proc. U.S. natn. Mus. 58(2349): 483-576. (text refs., key).

929. Muesebeck, C.F.W. 1970. The nearctic species of Orgilus Haliday (Hymenoptera: Braconidae). Smithson. Contribs Zool. 30. 104 pp. (53 refs., 57 figs., keys)

930. Muesebeck, C.F.W., K.V. Krombein, and H.K. Townes. 1951. Hymenoptera of America north of Mexico Synoptic Catalog. Agriculture Monogr.2. 1420 pp. [First supplement, 1958(305 pp). Second supplement, 1967 (584 pp.).] (text refs., 1 fig.)

931. Munroe, E. 1950. The occurrence of a butterfly in the Pribilof Islands. Lepid. News 4(4-5): 44. (text ref.)

932. Munroe, E.G. 1951. The geographic variation of Dasyuris polata (Duponchel) in North America (Lepidoptera, Geometridae). Can. Ent. 83(11): 290-294. (9 refs., 12 figs.)

933. Munroe, E.G. 1956a. The genus Orenaia in North America, with the description of a new species (Lepidoptera: Pyralidae). Can. Ent. 88(2): 74-78. (6 refs., 6 figs.)

934. Munroe, E. 1956b. Canada as an environment for insect life. Can. Ent. 88(7): 372-476. (269 refs., 61 figs., 1 tab.)

935. Munroe, E. 1960. The Pyralid fauna of the western United States. J. Lepid. Soc. 13[1959](4): 241-244. (0 ref.)

936. Munroe, E. 1963a. The gilvarius group of Aspilates Treitschke (Lepidoptera: Geometridae). Can. Ent. 95(3): 260-287. (21 refs., 56 figs., 1 key)

937. Munroe, E. 1963b. Systematic entomology survey program of the Canada Department of Agriculture. Proc. 9th Pacif. Sci. Congr.: 42-43. (0 ref.)

938. Munroe, E.G. 1963c. Characteristics and history of the North American fauna: Lepidoptera. Proc. 16th int. Congr. Zool. (Washington) 4: 21-27. (0 ref.)

939. Munroe, E. 1966. Revision of North American species of Udea Guenée (Lepidoptera: Pyralidae). Mem. ent. Soc. Can. 49. 57 pp. (40 refs., 92 figs., key)

940. Munroe, E. 1968. Insects of Ontario: geographical distribution and post-glacial origin. Proc. ent. Soc. Ont. 99: 43-50.(20 refs.)

941. Munroe, E. 1972. Pyraloidea, Pyralidae (part). In R.B. Dominick et al., The moths of America north of Mexico, Fasc. 13.1A. pp. 1-134. E.W. Classey Ltd., and R.D.B. Publications Inc., London, England. (text refs., figs. in other fasc., keys)

942. Munroe, E. 1973a. Pyraloidea, Pyralidae (part). In R.B. Dominick et al., The moths of America north of Mexico, Fasc. 13. 1C. pp. 253-304 + pls. E.W. Classey Ltd., and R.B.D. Publications Inc., London, England. (224 refs., 10 pls., 13 col. pls., keys)

943. Munroe, E. 1973b. A supposedly cosmopolitan insect: the celery webworm and allies, genus Nomophila Hübner (Lepidoptera: Pyralidae: Pyraustinae). Can. Ent. 105(2): 177-216. (47 refs., 47 figs.)

944. Munroe, E. 1976. Pyraloidea, Pyralidae (part). In R.B. Dominick et al., The moths of America north of Mexico. Fasc 13.2A. 102 pp. E.W. Classey Ltd., and R.B.D. Publications, Inc., London, England. (text refs., 4 colour pls., 8 pls., keys)

945. Munster, T. 1923. Report of the second Norwegian arctic expedition in the "Fram" 1898-1902. Coleoptera, Supplement. Norsk. ent. Tidsskr. 1(6): 297. (0 ref.)

946. Murdoch, J. 1885. pp. 133-135 in International Polar Expedition, 1882-1883. Rep. of the International Polar Expedition to Point Barrow, Alaska. [see Murdoch and Riley 1885]

947. Murdoch, J. and C.V. Riley. 1885. Part IV. Natural History, Sect. 4, Insects. in P.H. Ray et al., Report of the International Polar expedition to Point Barrow, Alaska. Government Printing Office, Washington. 695 pp. (0 ref.)

948. Muttkowski, R.A. 1915. Description of a trichopterous larva from the Pribilof Islands, Alaska. Bull. Wis. nat. Hist. Soc. 13(1): 42-45. (3 refs., 3 figs.)

949. Myers, B.J. 1959. Lice on Phoca hispida Schreber. Can. J. Zool. 37 (6): 1123. (0 ref.)

950. Nakahara, W. 1965. Contributions to the knowledge of the Hemerobiidae of western North America (Neuroptera). Proc. U.S. natn. Mus. 116(3503): 205-222, pl. 1 (27 refs., 4 figs., 1 pl., keys)

951. Nares, G.S. 1878. Narrative of a voyage to the polar sea during 1875-1876 in H.M. ships 'Alert' and 'Discovery'; with notes on the natural history, edited by H.W. Feilden, naturalist to the expedition. London, S. Lowe, Marston, Searle & Rivington. 2 vols. (xi, 395; viii, 378 pp.) (38 illus., 14 plates, incl. 7 mounted photos., 2 fold maps. Appendixes (vol. 2)) [see McLachlan 1878]

952. Natvig, L.R. 1930. Culiciden der 2 "fram" Expedition (1898-1902). Norsk ent. Tidsskr. 2(6): 358-359. (4 refs., 1 fig.)

953. Natvig, L.R. 1934. On some anomalies in Culicide hypopygiae. Norsk ent. Tidsskr. 3(5): 328-331. (4 refs., 4 figs.)

954. Natvig, L.R. 1945. Notes on Culex alpinus Linnaeus and Aedes nigripes (Zett.). Norsk ent. Tidsskr. 7(3-4): 99-106. (36 refs.)

955. Natvig, L.R. 1946. Differential characters of the female Aedes

nigripes (Zett.) and A. nearcticus Dyar. (Dipt. Cul.). Norsk ent. Tidsskr. 7(5): 164-167. (9 refs.)

956. Natvig, L.R. 1953. The type specimens of Aedes nigripes (Zett.). Norsk ent. Tidsskr. 9(1-2): 86-92. (11 refs., 4 figs.)

957. Nauman, J.W. and D.R. Kernodle. 1974. Aquatic organisms from selected sites along the proposed trans-Alaska pipeline corridor September 1970 to September 1972. U.S. Geol. Surv. Open-file Rep. 23 pp. (12 refs., 1 fig., 3 tabs.)

958. Nauman, J.W. and D.R. Kernodle. 1975. The effect of a fuel oil spill on benthic invertebrates and water quality on the Alaskan arctic slope, Happy Valley creek near Sagwon, Alaska. J. Res. U.S. Geol. Surv. 3(4): 495-500. (10 refs., 4 figs., 2 tabs.)

959. Nauman, J.W. and D.R. Kernodle. 1977. Aquatic organisms from selected sites along the trans-Alaska pipeline corridor, September 1970 to September 1972. U.S. Geol. Surv. Open-file Rep. 77-634. 55 pp. (Suppl.) (10 refs., 1 fig., 4 tabs.)

960. Nelson, B.C. 1972. A revision of the New World species of Ricinus (Mallophaga) occurring on Passeriformes (Aves). Univ. Calif. Publs Ent. 68. 175 pp. (text + 154 refs., 43 pls., 18 tabs., key)

961. Nettleship, D.N. 1973. Breeding ecology of turnstones Arenaria interpres at Hazen camp, Ellesmere Island, N.W.T. Ibis 115(2): 202-217. (42 refs., 7 figs., 2 tabs.)

962. Nettleship, D.N. 1974. The breeding of the Knot Calidris canutus at Hazen camp, Ellesmere Island, N.W.T. Polarforschung 44: 8-26. (44 refs., 9 figs., 3 tabs.)

963. Nielsen, E.T. 1963. [Mosquitoes in Greenland.] Grønland 1963: 295-301. (In Danish) (0 ref., 3 figs.)

964. Nielsen, E.T. and H.T. Nielsen. 1966. Observations on mosquitoes in Greenland. Meddr Grønland 170(3). 27 pp. (12 refs., 9 figs., 4 tabs.)

965. Nielsen, J.C. 1907. The insects of East Greenland. Meddr Grønland 29: 363-414. (text refs., 1 tab.) [Appendix by Becker, q.v.]

966. Nielsen, J.C. 1910. A catalogue of the insects of Northeast Greenland with descriptions of some larvae. Meddr Grønland 43: 55-68. (0 ref., 14 figs.) [and see Johansen 1910]

967. Nielsen, P. 1951. Limonia (Dicranomyia) vibei n. sp. from Grønland (Dipt., Tipul.). Report no. 3 from the zootopographical investigation of Grønland. Ent. Meddr 26(2): 185-186. (0 ref., 2 figs.)

968. Nimmo, A.P. 1971. The adult Rhyacophilidae and Limnephilidae (Trichoptera) of Alberta and eastern British Columbia and their post-glacial origin. Quaest. ent. 7(1): 3-234. (255 refs., 669 figs., 4 tabs., keys)

969. Nordenskiöld, N.A.E. 1882-1887. Vega expeditions Vetenskapliga iakttagelser Bearbetade af deltagare i resan och andra forskare. [Scientific observations of the Vega Expedition prepared by members of the expedition and other scientists.] Beijers,

Stockholm. 5 vols: 812, 516, 529, 582, 535 pp. (In Swedish and other languages) [see Aurivillius 1883, Kramer and Neumann 1883, McLachlan 1887, Sahlberg 1887]

970. Norton, D.W. 1971. Growth rates and bioenergetics of Calidridine sandpiper chicks of northern Alaska. Proc. 22nd Alaska Sci. Conf.: 16. [abstract only]

971. Norton, D.W., I.W. Ailes, and J.A. Curatolo. 1975. Ecological relationships of the inland tundra avifauna near Prudhoe Bay, Alaska. pp. 125-133 in J. Brown (Ed.), Ecological investigations of the tundra biome in the Prudhoe Bay region, Alaska. Univ. Alaska, Biol. pap. spec. rep. 2. 215 pp. (13 refs., 4 figs., 4 tabs.)

972. Nosek, J. 1977. A new genus and 6 new species of Protura from Alaska (Protura, Acerentomidae, Eosentomidae). Ent. Scand. 8 (4): 271-284. (6 refs., 7 figs.)

973. Nyquist, D. and J.D. LaPerrière. 1973. A preliminary survey of the zooplankton and benthos of an arctic lake near Prudhoe Bay, Alaska. Ent. News 84(7): 227-234. (11 refs., 9 figs.)

974. Obraztsov, N.S. 1959. Note on North American Aphelia species (Lepidoptera: Tortricidae). Am. Mus. Novit. 1964. 9 pp. (3 refs., 15 figs.)

975. Obraztsov, N.S. 1963. Some north American moths of the genus Acleris (Lepidoptera: Tortricidae). Proc. U.S. natn. Mus. 114(3469): 213-270. (28 refs., 7 + 60 figs., key)

976. Oliver, D.R. 1959. Some Diamesini (Chironomidae) from the nearctic and palaearctic. Ent. Tidskr. 80(1-2): 48-64. (16 refs., 19 figs.)

977. Oliver, D.R. 1961a. The Northern Insect Survey, 1957-1961. Arct. Circ. 14(4): 68-69. (0 ref.)

978. Oliver, D.R. 1961b. The insects of the Lake Hazen area, Ellesmere Island, 1961. Arct. Circ. 14(4): 69-71. (2 refs., 1 tab.)

979. Oliver, D.R. 1963. Entomological studies in the Lake Hazen area, Ellesmere Island, including lists of species of Arachnida, Collembola and Insecta. Arctic 16(3): 175-180. (9 refs.)

980. Oliver, D.R. 1964. A limnological investigation of a large arctic lake, Nettilling lake, Baffin Island. Arctic 17(2): 69-83. (30 refs., 1 fig., 5 tabs.)

981. Oliver, D.R. 1968a. Adaptations of arctic Chironomidae. Ann. zool. fenn. 5(1): 111-118. (17 refs., 4 figs., 2 tabs.)

982. Oliver, D.R. 1968b. Insects. pp. 416-436 in C.S. Beals and D.A. Shenstone (Eds.), Science, History and Hudson Bay, Vol. I. Energy, Mines and Resources, Ottawa. 501 pp. (37 refs., 7 figs.)

983. Oliver, D.R. 1970. Designation and description of lectotypes of the 6 Greenland Orthocladiinae (Diptera, Chironomidae) described by Lundbeck in 1898. Ent. Scand. 1(2): 102-108. (11 refs., 18 figs., text tabs.)

984. Oliver, D.R. 1976. Chironomidae (Diptera) of Char Lake, Cornwallis Island, N.W.T., with descriptions of two new species. Can. Ent.

108(10): 1053-1064. (29 refs., 24 figs., 2 tabs.)

985. Oliver, D.R. 1977. Bicinctus-group of the genus Cricotopus Van der Wulp (Diptera: Chironomidae) in the Nearctic with a description of a new species. J. Fish. Res. Bd Can. 34(1): 98-104. (9 refs., 8 figs., 2 tabs.)

986. Oliver, D.R. and P.S. Corbet. 1966. Aquatic habitats in a high arctic locality: the Hazen Camp study area, Ellesmere Island, N.W.T. Def. Res. Bd Can., D. Phys.R.(G), Hazen 26. 115 pp. + 267 figs. (41 refs., 267 figs., 14 tabs.)

987. Oliver, D.R., P.S. Corbet, and J.A. Downes. 1964. Studies on arctic insects: the Lake Hazen project. Can. Ent. 96(1-2): 138-139. (0 ref.)

988. Oliver, D.R. and H.V. Danks. 1972. Sex ratios of some high arctic Chironomidae (Diptera). Can. Ent. 104(9): 1413-1417. (16 refs., 2 tabs.)

989. Oliver, D.R., D. McClymont, and M.E. Roussel. 1978. A key to some larvae of Chironomidae (Diptera) from the Mackenzie and Porcupine River watersheds. Can. Fish. Mar. Serv. Tech. Rep. 791. 73 pp. (54 refs., 207 figs., 2 tabs., keys)

990. Omer-Cooper, J. 1931. Gyrinus opacus var. lecontei O.-C. (nec Fall, 1921) = G. opacus var. blairi nom. nov. Entomologist's mon. Mag. 67: 239. (text refs.)

991. O'Neill, K. and R.S. Bigelow. 1964. The Taeniothrips of Canada (Thysanoptera: Thripidae). Can. Ent. 96(9): 1219-1239. (53 refs., 44 figs., 1 tab., key)

992. O'Neill, M.C.A. and A.G. Robinson. 1977. Ant-aphid associations in the province of Manitoba. Manitoba Ent. 11: 74-88. (16 refs., 4 tabs.)

993. Opell, B.D. and J.A. Beatty. 1976. The nearctic Hahniidae (Arachnida: Araneae). Bull. Mus. comp. Zool. Harv. 147(9): 393-433. (22 refs., 114 figs., 8 maps, keys)

994. Orchymont, A.d' and W.J. Brown. 1940. Helophorus arcticus Brown, a living fossil (Coleoptera). Can. Ent. 72(1): 1-4. (text refs., 2 figs.)

995. Osten Sacken, C.R. 1878. Report on the Diptera brought home by Dr. Bessels from the Arctic voyage of the "Polaris" in 1872. Proc. Boston Soc. nat. Hist. 19: 41-43. (text refs.)

996. Osten Sacken, C.R. 1879. Diptera. pp. 116-118 in R. McLachlan, Report on the Insecta (including Arachnida) collected by Captain Feilden and Mr. Hart between the parallels of 78° and 83° North Latitude during the recent Arctic Expedition. J. Linn. Soc. (Zool.) 14: 98-122. (0 ref.)

997. Oughton, J.G. 1939. The burrow of an arctic spider, Lycosa asivak Emerton. Can. Fld Nat. 53(8): 123-124. (text refs.)

998. Packard, A.S. 1877a. Exploration of the Polaris expedition to the North Pole. Am. Nat. 11: 51-53. (0 ref.)

999. Packard, A.S. 1877b. The insects of the American ("Polaris") Expedition. Entomologist's mon. Mag. 13: 228-229. (0 ref.) [same as

Packard 1877a.]

1000. Panfilov, D.V., V.F. Shamurin, and B.A. Yurtsev. 1960. [The conjugate distribution of bumblebees and legumes in the arctic.] Byull. Mosk. Obshch. Ispyt. Prir. Nov. Ser., Otdel. Biol. 65(3): 53-62. (In Russian, English summary) (22 refs., 4 figs., 1 tab.)

1001. Panfilov, D.V., V.F. Shamurin, and B.A. Yurtsev. 1962. [More on the conjugate distribution of bumblebees and legumes in the arctic.] Byull. Mosk. Obshch. Ispyt. Prir. Nov. Ser., Otdel. Biol. 67(3): 130-131. (In Russian) (3 refs.)

1002. Parmelee, D.F. and S.D. MacDonald. 1960. The birds of west-central Ellesmere Island and adjacent areas. Bull. natn. Mus. Can. 169: 1-103. (66 refs., 2 figs., 10 pls., text tabs.)

1003. Parry, W.E. 1821-1824. Voyage for the discovery of a northwest passage, in the years 1819-1820, containing an account of the subjects of natural history. 310 pp. + appendix clxxix pp. (1821); Supplement to the Appendix, pp. clxxx-cccx (1824) Murray, London. [see Kirby 1824]

1004. Parry, W.E. 1826. Journal of a third voyage for the discovery of a north-west passage from the Atlantic to the Pacific; performed in the years 1824-1825 in his majesty's ships Hecla and Fury, under the orders of Captain William Edward Parry, R.N., F.R.S. Murray, London. [see Ross 1826]

1005. Parry, W.E. 1828. Narrative of an attempt to reach the North Pole in boats fitted for the purpose, and attached to his Majesty's ship Hecla, in the year 1827. Murray, London. 229 pp.

1006. Patch, E.M. 1923. Homoptera. pp. 143-144 in W.L. McAtee (Ed.), A biological survey of the Pribilof Islands, Alaska. Part II, Insects, arachnids and chilopods of the Pribilof Islands, Alaska. N. Am. Fauna 46: 129-255. (text refs., 7 figs.)

1007. Pattie, D.L. 1977. Population levels and bioenergetics of arctic birds on Truelove lowland. pp. 413-436 in L.C. Bliss (Ed.), Truelove lowland, Devon Island, Canada: A high arctic ecosystem. Univ. Alta Press, Edmonton. 714 pp. (51 refs., 4 figs., 7 tabs.)

1008. Payne, F.F. 1887. Flora and fauna of Prince of Wales Sound, Hudson Straits. pp. 70-83 in Report of the Hudson's Bay Expedition of 1886 under the command of Lieut. A.R. Gordon, R.N. Canada Marine and Fisheries Dept. [insects p. 79] (0 ref.)

1009. Pedersen, B.V. 1978. Studies on the taxonomy of Chironomus islandicus (Kieffer, 1913) (Diptera: Chironomidae). Ent. Scand. 9(4): 309-311. (6 refs., 3 figs., 1 tab.)

1010. Peters, H.B. 1934. Liste des bisher bestimmten Materials. p. 174 in H.B. Peters et al., Wissenschaftliche Ergebnisse der Deutschen Grönland-expedition Alfred Wegener 1929 und 1930/31. Vol. II. Anthropologie und Zoologie. 196 pp. (0 ref.)

1011. Peters, H.S. 1934. V. Mallophaga from birds of Southampton Island, Hudson Bay. The exploration of Southampton Island, Hudson Bay, by G.M. Sutton. Mem. Carnegie Mus. 12, part II, sect. 4: 35-37. (0 ref.)

1012. Peterson, B.V. 1970. The Prosimulium of Canada and Alaska (Diptera:

Simuliidae). Mem. ent. Soc. Can. 69. 216 pp. (123 refs., 158 figs., 15 maps., 1 tab., keys)

1013. Peterson, P.C. 1972. The feather mite family Alloptidae Gaud. 3 - The Echinocarinae, new subfamily (Acarina, Analgoidea). Steenstrupia (Copenhagen). 2(14): 197-205. (15 refs., 8 figs., 1 tab.)

1014. Petrunkevitch, A. 1911. A synonymic index-catalogue of spiders of North, Central and South America, with all adjacent islands, Greenland, Bermuda, West Indies, Tierra del Fuego, Galapagos, etc. Bull. Am. Mus. nat. Hist. 29: 1-793. (533 refs.)

1015. Philip, C.B. 1947. A catalog of the blood-sucking fly family Tabanidae (horseflies and deerflies) of the Nearctic Region north of Mexico. Am. midl. Nat. 37(2): 257-324. (text refs., index, key)

1016. Philip, K.W. 1970. North American annual summary for 1969. Zone VIII: The far north. News Lepid. Soc. 1970(3): 16-18. (0 ref.)

1017. Philip, K.W. 1971a. Composition and abundance of the Lepidopteran fauna of arctic and alpine Alaskan tundra. pp. 133-135 in J. Brown and S. Bowen (Eds.), The structure and function of the tundra ecosystem, Vol. I. U.S. Tundra Biome Program, U.S. International Biological Program. 282 pp. (1 ref.)

1018. Philip, K.W. 1971b. Butterflies of the tundra. Proc. 22nd Alaska Sci. Conf: 31. [abstract only]

1019. Philip, K.W. 1971c. Annual field season summary for 1970. Zone 8: The far north. News Lepid. Soc. 1971(2): 5-8. (0 ref.)

1020. Philip, K.W. 1972. The 1971 field season summary. Zone 8: The far north. News Lepid. Soc. 1972(2): 14-15. (0 ref.)

1021. Philip, K.W. 1973. 1972 field season summary. Zone 8: The far north. News Lepid. Soc. 1973(2): 17-18. (0 ref.)

1022. Philip, K.W. 1974. The 1973 field season summary. Zone 8: The far north. News Lepid. Soc. 1974(2): 14-16. (0 ref.)

1023. Philip, K.W. 1975. The 1974 field season summary. Zone 8: The far north. News Lepid. Soc. 1975(2/3): 15-17. (0 ref.)

1024. Philip, K.W. 1976. 1975 field season summary, concluded. Zone 8: The far north. News Lepid. Soc. 1976(3): 1-5. (0 ref.)

1025. Philip, K.W. 1977. Supplement to the field summary [1976]. Zone 8: The far north. News Lepid. Soc. 1977(4): 1-3. (0 ref.)

1026. Philip, K.W. 1978. Field summary for 1977. Zone 8. The far north. News Lepid. Soc. 1978(2): 15. (0 ref.)

1027. Pickard Cambridge, O. 1877. On some new and little-known spiders from the arctic regions. Ann. Mag. nat. Hist. 4(20): 273-285. (text refs., 7 figs.)

1028. Pienkowski, M.W. 1977. Studies of the feeding and breeding behaviour of ringed plovers in the Mestersvig area. pp. 69-70 in G.H. Green and J.J.D. Greenwood (Eds.), Report of the Joint Biological Expedition to N.E. Greenland 1974. Dundee: Dundee

University N.E. Greenland Expedition. 258 pp. (0 ref.)

1029. Pitelka, F.A. 1959. Numbers, breeding schedule, and territoriality in Pectoral Sandpipers of northern Alaska. Condor 61: (4): 233-264. (38 refs., 7 figs., 3 tabs.)

1030. Pittard, D. and R.W. Strandtmann. 1971. A comparison of a minuscule part of the fauna of northern Alaska with a similar fauna of Antarctica. Texas J. Sci. 22(2-3): 298. [abstract only]

1031. Platnick, N.I. and M.U. Shadab. 1975a. A revision of the spider genus Gnaphosa (Araneae, Gnaphosidae) in America. Bull. Am. Mus. nat. Hist. 155(1): 1-66. (55 refs., 150 figs., 15 maps, key)

1032. Platnick, N.I. and M.U. Shadab. 1975b. A revision of the spider genera Haplodrassus and Orodrassus (Araneae, Gnaphosidae) in North America. Am. Mus. Novit. 2583. 40 pp. (32 refs., 106 figs., 9 maps, key)

1033. Polar Record. 1953. Studies of blood-sucking flies in northern Canada, 1947-1951. Polar Rec. 6(4): 675-679. (text refs.)

1034. Poppius, B.R. 1910. Die Coleopteren des arktischen Gebietes. Fauna arct. 5: 289-447. (160 refs.)

1035. Poulton, E.B. 1934a. Lepidoptera. Oxford University Hudson Strait Expedition to Akpatok Island, Ungava Bay, 1931. Ann. Mag. nat. Hist. 10(14): 48-50. (0 ref.)

1036. Poulton, E.B. 1934b. Hymenoptera parasitica, including a few specimens taken at Port Burwell, Labrador. Oxford University Hudson Strait Expedition to Akpatok Island, Ungava Bay, 1931. Ann. Mag. nat. Hist. 10(14): 50-53. (0 ref.)

1037. Power, G. 1969. The salmon of Ungava Bay. Arct. Inst. N. Am. Tech. Pap. 22. 72 pp. (68 refs., 20 figs., 24 tabs. [diet, pp. 51-55]

1038. Preble, E.A., W.L. McAtee, et al. 1923. A biological survey of the Pribilof Islands, Alaska. N. Am. Fauna 46. 255 pp. (U.S. Biol. Surv. U.S. Govt. printing off.) [see McAtee 1923]

1039. Procter, D.L.C. 1977. Invertebrate respiration on Truelove lowland. pp. 383-393 in L.C. Bliss (Ed.), Truelove lowland, Devon Island, Canada: A high arctic ecosystem. Univ. Alta Press, Edmonton. 714 pp. (24 refs., 1 fig., 5 tabs.)

1040. Quay, W.B. 1949. Further description of Polyplax alaskensis Ewing (Anoplura). Psyche, Camb. 56(4): 180-183. (text refs., 2 figs.)

1041. Quervain, A. De and P.L. Mercanton. 1920. Ergebnisse der Schweizerischen Grönlandexpedition, 1912-1913. Schweiz. Naturforsch. Ges. Denkschrift 53: 229-307. [see Mercanton 1920]

1042. Quervain, A. De and P.L. Mercanton. 1925. Résultats scientifiques de l'expédition suisse au Groenland 1912-1913. Meddr Grønland 59: 55-271. [see Mercanton 1925]

1043. Ramme, W. 1928. Orthopteren des arktischen Gebietes. Fauna arct. 5: 639-646. (10 refs.)

1044. Rausch, R.L. and V.R. Rausch. 1971. The somatic chromosomes of some north American marmots (Sciuridae), with remarks on the relationships of Marmota broweri Hall and Gilmore. Mammalia 35(1): 85-101. (43 refs., 6 figs.)

1045. Ray, P.H. et al. 1885. Report of the International Polar expedition to Point Barrow, Alaska. Part IV, Natural History, Sect. 4: Insects by J. Murdoch and C.V. Riley; Sect. 5: Marine Invertebrates by J. Murdoch: Sect 6: Report on the mollusks by W.H. Dall. U.S. Govt. printing Office, Washington. 695 pp. [see Murdoch and Riley 1885]

1046. Razowski, J. 1966. World fauna of the Tortricini (Lepidoptera, Tortricidae). Zaklad. Zool. Syst. w. Krakowie, Polsk. Akad. Nauk. 576 pp. + XLI pls. (201 refs., 832 figs., 41 pls., keys)

1047. Rebel, H. 1907. Bericht der Sektion fur Lepidopterologie II. [untitled] [Greenland Lepidoptera collected by Dr. Stiasny in West Greenland]. Ver. zool. bot. Ges. Wien. 57: 28-30. (6 refs.)

1048. Rebel, H. 1911. Versammlung am 9 December 1910. IV. [untitled] [Lepidoptera from East Greenland]. Ver. zool. bot. Ges. Wien. 61(1-2): 43-44 (0 ref.)

1049. Reiche, L. 1857. Description sommaire de cinq espèces nouvelles d'Insectes provenant de L'expédition aux Mers Arctiques, effectuée en 1856 sous la direction de S.A.I. le prince Napoléon. Ann. Soc. ent. Fr. (3) V.Bull. VIII-X. (0 ref.)

1050. Rempel, J.G. 1950. A guide to the mosquito larvae of western Canada. Can. J. Res. (D) 28(4): 207-248. (15 refs., 43 figs., 1 map, keys)

1051. Remy, P. 1928a. Les Collemboles du Groenland. Meddr Grønland 74: 55-70. (16 refs.)

1052. Remy, P. 1928b. Arthropodes terrestres récoltés au Groenland au cours de la croisière du "Pourquoi-Pas?" en 1926. Bull. Soc. linn. Lyon. 7: 51-53. (0 ref.)

1053. Remy, P. 1928c. Quelques arthropodes récoltés dans la région du Scoresby-Sund (Groenland Oriental) au cours de la croisière arctique du "Pourquoi-pas?" en 1926. Meddr Grønland 74: 85-95. (17 refs., 1 fig.)

1054. Richard, P.W. and F.C. Harmston. 1972. Diptera and other arthropods of the Sukkertoppen Tasersiaq area, southwest Greenland. Arctic 25(2): 107-114. (7 refs., 4 figs., 6 tabs.)

1055. Richards, K.W. 1970. Biological studies of arctic bumblebees. Unpub. M.Sc. thesis, Univ. Alta., Edmonton. 165 pp.

1056. Richards, K.W. 1973. Biology of Bombus polaris Curtis and B. hyperboreus Schönherr at Lake Hazen, Northwest Territories (Hymenoptera: Bombini). Quaest. ent. 9(2): 115-157. (126 refs., 29 figs., 10 tabs.)

1057. Richards, O.W. 1927. Some notes on the bumble-bees allied to Bombus alpinus, L. Tromsø Mus. Årsh. 50(6). 32 pp., II pl. [Issued also as Oxford University Exploration Club, Greenland and Spitsbergen papers no. 14] (73 refs., 2 pls., keys)

1058. Richards, W.R. 1961. Two new species of aphids from eastern boreal and arctic Canada (Homoptera: Aphididae). Can. Ent. 93(7):

497-501. (4 refs., 8 figs.)

1059. Richards, W.R. 1963. The Aphididae of the Canadian Arctic (Homoptera). Can. Ent. 95(5): 449-464. (13 refs., 2 figs.)

1060. Richards, W.R. 1964a. A new arctic aphid (Homoptera: Aphididae). Can. Ent. 96(7): 1027-1029. (1 ref., 2 figs.)

1061. Richards, W.F. 1964b. The scale insects of the Canadian Arctic (Homoptera: Coccoidea). Can. Ent. 96(11): 1457-1462. (5 refs., 16 figs.)

1062. Richards, W.R. 1972. Acyrthosiphon pedicularis, a new aphid from the District of Keewatin, with a key to the Canadian species (Homoptera: Aphididae). Can. Ent. 104(8): 1149-1152. (0 ref., 2 figs., key)

1063. Richardson, J. 1829-1837. Fauna Boreali-Americana: containing descriptions of the objects of natural history collected on the late northern land expeditions, under command of Captain Sir John Franklin, R.N. 4 vols. [see Kirby 1837]

1064. Richardson, J. 1851. Arctic Searching Expedition: a journal of a boat voyage through Rupert's Land and the arctic sea in search of the Discovery ships under command of Sir John Franklin. 2 vols. [see White 1851]

1065. Richter, W. 1928. Die Thysanopteren des arktischen Gebietes. Fauna arct. 5: 835-850. (24 refs., keys)

1066. Ricker, W.E. 1938. A new stonefly from Baffin Land (Plecoptera, Capniidae). Can. Ent. 70(8): 173-174. (0 ref., 1 fig.)

1067. Ricker, W.E. 1944. Some Plecoptera from the far north. Can. Ent. 76(9): 174-185. (13 refs., 22 figs., key)

1068. Ricker, W.E. 1952. Systematic studies in Plecoptera. Indiana Univ. Publ. Sci. Ser. 18. 200 pp. (14 refs., 154 figs., keys)

1069. Ricker, W.E. 1964a. Distribution of Canadian stoneflies. Fish. Res. Bd Can. Manuscript Rep. Ser. (Biol.) 770. 24 pp. + 40 figs. (0 ref., 40 figs.)

1070. Ricker, W.E. 1964b. Distribution of Canadian stoneflies. Gewässer Abwässer. 34/35: 50-71. (13 refs., 20 figs.)

1071. Ricker, W.E. 1965. New records and descriptions of Plecoptera (Class Insecta). J. Fish. Res. Bd Can. 22(2): 475-501. (29 refs., 54 figs., key)

1072. Ricker, W.E., R. Malouin, P. Harper, and H.H. Ross. 1968. Distribution of Quebec stoneflies (Plecoptera). Naturaliste can. 95(5): 1085-1123. (32 refs.)

1073. Rigler, F.H. 1972. The Char lake project. A study of energy flow in a high arctic lake. pp. 287-300 in Z. Kajak and A. Hillbricht-Ilkowska (Eds.), Productivity problems of freshwaters. (Proc. Int. Biol. Prog. UNESCO Symposium Kazimierz Dolny, Poland May 6-12, 1970.) 918 pp. Polish Scientific Publishers, Warsaw. (17 refs., 5 figs., 3 tabs.)

1074. Rigler, F.H. 1974. Char Lake Project PF-2. Final Report. Can. Committee for Int. Biol. Programme. 96 pp. (4 refs., 15 figs., 6 tabs, 12 appendixes) [see Rigler 1975]

1075. Rigler, F.H. 1975. Chap. 10. The Char Lake Project. An introduction to limnology in the Canadian arctic. pp. 171-198 in T.W. M. Cameron and L.W. Billingsley (Eds.), Energy flow - its biological dimensions. A summary of the IBP in Canada: 1964-1974. Royal Society of Canada, Ottawa. 319 pp. (33 refs., 11 figs., 1 tab.)

1076. Riley, C.V. 1885. Report on a collection of insects made at Point Barrow, Alaska. pp. 134-135 in P.H. Ray et al., Report of the International Polar expedition to Point Barrow, Alaska. Government Printing office, Washington. 695 pp. [see Murdoch and Riley 1885]

1077. Ringdahl, O. 1933. Tachiniden und Musciden aus Nordost-Grönland. (Zoological results of the Norwegian Scientific Expeditions to East Greenland. II). Skr. Svalbard Ishavet 53: 15-18. (text refs.)

1078. Roback, S.S. 1971. The adults of the subfamily Tanypodinae (=Pelopiinae) in North America (Diptera: Chironomidae). Monogr. Acad. nat. Sci. Philad. 17. 410 pp. (158 refs., 3 + 615 figs., 36 maps, 33 tabs., keys)

1079. Robinson, A.G. 1979. Annotated list of aphids (Homoptera: Aphididae) collected at Churchill, Manitoba with descriptions of new species. Can. Ent. 111(4): 447-458. (9 refs., 4 figs.)

1080. Robinson, A.G. and V. Rojanavongse. 1976. A key to the nearctic species of Aphis L. (Homoptera: Aphididae) on Ribes spp., with descriptions of two new species from Manitoba. Can. Ent. 108 (2): 159-164. (12 refs., 24 figs., key)

1081. Rohwer, S.A. 1920. Descriptions of twenty-five new species of north American Hymenoptera. Proc. U.S. natn. Mus. 57(2312): 209-231. (text refs., key)

1082. Rojanavongse, V. and A.G. Robinson. 1977. Species of Aphis (Homoptera:Aphididae) in Manitoba, with a key, and descriptions of new species. Can. Ent. 109(5): 649-661. (27 refs., 40 figs., key)

1083. Roland, J. 1978. Variation in spectral reflectance of alpine and arctic Colias (Lepidoptera: Pieridae). Can. J. Zool. 56(6): 1447-1453. (23 refs., 3 figs., 4 tabs.)

1084. Roman, A. 1916. Ichneumoniden aus West-Grönland. Ark. Zool.10(22). 12 pp. (text refs.)

1085. Roman, A. 1930. Oxford University Greenland Expedition, 1928. Ichneumonidae collected by Maj. R.W.G. Hingston on the Oxford University Expedition to Greenland, 1928. Ann. Mag. nat. Hist. 10(5): 281-288. (text refs.)

1086. Roman, A. 1933. Schlüpfwespen aus Ost-Grönland. (Zoological results of the Scientific Expeditions to East-Greenland. II). Skr. Svalbard Ishavet 53: 5-13. (0 ref.)

1087. Roman, A. 1934. East Greenland ichneumonids. Ann. Mag. nat. Hist. 10(14): 606-611. (0 ref.)

1088. Romer, A. (Ed.). 1900-1932. Fauna Arctica. 6 vols. [see Arndt 1931,

Friese 1902, Jordan 1932, Kiaer 1902, Meijere 1910, Poppius 1910, Ramme 1928, Richter 1928, Schäffer 1900, Stitz 1928, Stitz and Ramme 1928, Strand 1906, Trägårdh 1904, Ulmer 1932, Viets 1931, Zunker 1932]

1089. Rosenberg, D.M., H.V. Danks, J.A. Downes, A.P. Nimmo, and G.E. Ball. 1979. Procedures for a faunal inventory. pp. 509-532 in H.V. Danks (Ed.), Canada and its insect fauna. Mem. ent. Soc. Can. 108. 573 pp. (71 refs., 1 fig., 1 tab.)

1090. Ross, H.H. 1941. Descriptions and records of North American Trichoptera. Trans. Am. ent. Soc. 67(1-2): 35-126. (text refs., 98 figs., keys)

1091. Ross, H.H. and D.R. Merkley. 1952. An annotated key to the nearctic males of Limnephilus (Trichoptera, Limnephilidae). Am. midl. Nat. 47(2): 435-455. (21 refs., 25 figs., key)

1092. Ross, H.H. and J.C. Morse. 1973. Micrasema kluane, a probable stepping stone to the arctic (Trichoptera, Brachycentridae). Ent. News 84(9): 291-293. (5 refs., 2 figs.)

1093. Ross, J.C. 1826. Zoology. Appendix pp. 91-120 in W.E. Parry, Journal of a third voyage for the discovery of a north-west passage from the Atlantic to the Pacific; performed in the years 1824-1825 in his majesty's ships Hecla and Fury, under the orders of Captain William Edward Parry, R.N.,F.R.S. Murray, London. (text refs.)

1094. Ross, J.F. and S.M. Smith. 1976. A review of the mermithid parasites (Nematoda: Mermithidae) described from North American mosquitoes (Diptera: Culicidae) with descriptions of three new species. Can. J. Zool. 54(7): 1084-1102. (28 refs., 28 figs., key)

1095. Rosse, I.C. 1883. Cruise of the Revenue-Steamer Corwin in Alaska and the N.W. Arctic Ocean in 1881. Notes and memoranda: medical and anthropological; botanical; ornithological. 120 pp. U.S. Govt. printing office, Washington. (0 ref.) [p. 24]

1096. Rosswall, T. and O.W. Heal (Eds.). 1975. Structure and function of tundra ecosystems. Ecol. Bull. (Stockholm) 20. 450 pp. [see Bliss 1975b; Bunnell et al. 1975]

1097. Rostrup, S. 1901. Gronlandske phytoptider. Vidensk. Meddr dansk naturh. Foren. 52: 241-249., 1 pl. (In Danish) (text refs., 1 pl.)

1098. Roth, V.D. 1952. A review of the genus Tegenaria in North America (Arachnida: Agelenidae). J. Wash. Acad. Sci. 42(9): 283-288. (24 refs., 5 figs., key)

1099. Roth, V.D. 1968. The spider genus Tegenaria in the western hemisphere (Agelenidae). Am. Mus. Novit. 2323. 33 pp. (31 refs., 39 figs., key)

1100. Roth, V.D. and P.L. Brame. 1972. Nearctic genera of the spider family Agelenidae (Arachnida, Araneida). Am. Mus. Novit. 2505. 52 pp. (44 refs., 70 figs., keys)

1101. Rubsaamen, E.H. 1898. Grönländische Mycetophiliden, Sciariden, Cecidomyiden, Psylliden, Aphiden und Gallen. Zoologische Ergebnisse der Gesellschaft fur Erdkunde zu Berlin unter Leitung Dr. von Drygalski aus gesandten Grönlandexpedition nach Dr. Van Vanhöffen's Sammlung bearbeitet, VIII. Bibliotheca Zoologica

20: 103-119, 2 pls. (text refs., 11 figs., 2 pls., key)

1102. Ryan, J.K. 1972. Devon Island invertebrate research. pp. 293-314 in L.C. Bliss (Ed.), Devon Island IBP Project, High arctic ecosystem. Project report 1970 and 1971. Univ. Alta., Edmonton. 413 pp. (48 refs., 6 figs., 4 tabs.) [superceded, see Ryan 1977a, 1977b.]

1103. Ryan, J.K. 1976. Insects in the high Arctic. Proc. 15th int. Congr. Ent. (Washington 1976): 167-170. (25 refs., 3 tabs.)

1104. Ryan, J.K. 1977a. Synthesis of energy flows and population dynamics of Truelove Lowland invertebrates. pp. 325-346 in L.C. Bliss (Ed.), Truelove lowland, Devon Island, Canada: A high arctic ecosystem. Univ. Alta Press, Edmonton. 714 pp. (76 refs., 1 fig., 4 tabs.)

1105. Ryan, J.K. 1977b. Appendix 7. Invertebrates of Truelove lowland. pp. 699-703 in L.C. Bliss (Ed.), Truelove lowland, Devon Island, Canada: A high arctic ecosystem. Univ. Alta. Press, Edmonton. 714 pp.

1106. Ryan, J.K. 1977c. Energy flow through arctic invertebrates at Truelove lowland, Devon Island, N.W.T. Unpub. Ph.D. Thesis, Univ. Alta. 239 pp.

1107. Ryan, J.K. manuscript. Invertebrate faunas at I.B.P. tundra sites. [in L.C. Bliss, J.B. Cragg, O.W. Heal, and J.J. Moore(Eds.), Tundra ecosystems. A comparative analysis of ecosystems. Cambridge Univ. Press,Cambridge](64 refs., 6 tabs.) [pub. 1981]

1108. Ryan, J.K. and C.R. Hergert. 1977. Energy budget for Gynaephora groenlandica (Homeyer) and G. rossii (Curtis) (Lepidoptera: Lymantriidae) on Truelove Lowland. pp. 395-409 in L.C. Bliss (Ed.), Truelove lowland, Devon Island, Canada: A high arctic ecosystem. Univ. Alta Press, Edmonton. 714 pp. (34 refs., 2 figs., 5 tabs.)

1109. Sabrosky, C.W. 1943. A revised synopsis of nearctic Thaumatomyia (=Chloropisca) (Diptera, Chloropidae). Can. Ent. 75(6): 109-117 (text refs., key)

1110. Sabrosky, C.W. 1958. New species and notes on North American acalyptrate Diptera. Ent. News 69(7): 169-176. (text refs.)

1111. Sabrosky, C.W. 1959. The nearctic species of the filth fly genus Meoneura (Diptera, Milichiidae). Ann. ent. Soc. Am. 52(1): 17-26. (text refs., 12 figs., key)

1112. Saether, O.A. 1970. Nearctic and Palaearctic Chaoborus (Diptera: Chaoboridae). Bull. Fish. Res. Bd Can. 174. 57 pp. (56 refs., text + 1 tab., 19 figs.)

1113. Saether, O.A. 1975. Nearctic and Palaearctic Heterotrissocladius (Diptera: Chironomidae). Bull. Fish. Res. Bd Can. 193. 67 pp. (114 refs., 15 figs., keys)

1114. Saether, O.A. 1976. Revision of Hydrobaenus, Trissocladius, Zalutschia, Paratrissocladius, and some related genera (Diptera: Chironomidae). Bull. Fish. Res. Bd Can. 195. 287 pp. (157 refs., 84 figs., keys)

1115 Saether, O.A. 1977. Taxonomic studies on Chironomidae: Nanocladius,

Pseudochironomus, and the Harnischia complex. Bull. Fish. Res. Bd Can. 196. 143 pp. (117 refs., 49 figs., keys)

1116. Sahlberg, J. 1887. [Coleoptera and Hemiptera, collected by members of the Vega expedition on the American coast of Bering Strait, in the vicinity of Port Clarence, Grantley Harbor and Lake Iman-Ruk, July 23-26, 1879.] pp. 43-57 in N.A.E. Nordenskiöld (Ed.), Vega expeditionens vetenskapliga iakttageleser. (Vol. 4). (In Swedish) (text ref., 1 tab.)

1117. Sahlberg, J.R. 1901a. [Aleocharids collected in arctic regions by the Swedish expeditions of 1883 [i.e. 1898] and 1899.] Ent. Tidskr. 22(2-3): 167-169. (In Swedish) (text refs.)

1118. Sahlberg, J.R. 1901b. [Coleoptera in polar regions.] Mem. Soc. Fauna Flora fenn. 27: 96-98, 177. (0 ref.) [177: summary in German]

1119. Sailer, R.I. 1951. Concerning a recent paper on the insects and related arthropods of arctic Alaska. Ecology 32(4): 729-730. (text refs.)

1120. Sailer, R.I. 1952. Circumpolar distribution of water boatmen (Hemiptera: Corixidae). Can. Ent. 84(9): 280. (1 ref.)

1121. Sailer, R.I. 1953. The blackfly problem in Alaska. Mosquito News 13(4): 232-235. (1 ref.)

1122. Sailer, R.I. 1955. Invertebrate research in Alaska. Arctic 7(3-4): 266-274. [also pp. 154-162 in Arctic Research. Arct. Inst. N. Am. Spec. Publ. 2(1955)]. (62 refs.)

1123. Sailer, R.I. and S.E. Lienk. 1951. Blowflies (Calliphoridae) in Alaska. Can. Ent. 83(8): 208-211. (3 refs., 1 tab.)

1124. Sailer, R.I., A.V. Regnier, Jr., and J.U. McGuire, Jr. 1955. Use of precipitation data for predicting mosquito abundance in Alaska. Final Rep., Arct. Inst. N. Am.,project ONR-95. 72 pp. (0 ref., 14 figs., 2 charts, 7 tabs.)

1125. Sanderson, M.W. 1946. Nearctic Stenus of the croceatus group. (Coleoptera, Staphylinidae). Ann. ent. Soc. Am. 39(3): 425-430. (4 refs., 22 figs., key)

1126. Savile, D.B.O. 1964. Review of A. Löve and D. Löve (Eds.), North Atlantic biota and their history. Arctic 17(2): 137-141. (0 ref.)

1127. Säwedal, L. 1979. Micropsectra brundini n. sp. (Diptera: Chironomidae) and geographical variation in the Micropsectra insignilobus aggregate. Ent. Scand. Suppl. 10: 133-138. (9 refs., 1 fig., 2 tabs.,key)

1128. Scanlon, J.E. and P.T. Johnson. 1957. On some microtine-infesting Polyplax (Anoplura). Proc. ent. Soc. Wash. 59(6): 279-283. (14 refs., 11 figs., key)

1129. Schäffer, C. 1900. Die arktischen und subarktischen Collembola. Fauna arct. 1: 235-258. (51 refs., 1 tab.)

1130. Scheiring, J.R. 1977. Geographic variation in Scatella stagnalis (Diptera: Ephydridae). Ann. ent. Soc. Am. 70(4): 511-523. (24 refs., 9 figs., 3 tabs.)

1131. Schiödte, J.C. 1857. [Survey of the land, freshwater and shore arthropods of Greenland.] Tillaeg til H. Rink: Grønland geografisk og statistisk beskrevet 2, 50-74. Kobenhaven. (In Danish) [for translation into German see Schiödte 1859]

1132. Schiödte, J.C. 1859. Ubersicht der Land-, Susswasser- und Ufer-Arthropoden Grönlands. Berl. ent. Z. 3: 134-157. [translated from Danish (Schiödte 1857) by A.V. Etzel] (text refs.)

1133. Schmid, F. 1952. Le groupe de Chilostigma (Trichopt., Limnoph.). Arch. Hydrobiol. 47(1): 75-163. (text refs., 198 figs.)

1134. Schmid, F. 1954a. Contribution à l'étude de la sous-famille des Apataniinae (Trichoptera, Limnophilidae). II. Tijdschr. Ent. 97(1-2): 1-74. (72 refs., 100 figs.)

1135. Schmid, F. 1954b. Le genre Asynarchus McL. (Trichopt., Limnoph.). Mitt.Schweiz ent. Ges. 27(2): 57-96. (text refs., 27 figs.)

1136. Schmid, F. 1964. Some nearctic species of Grammotaulius Kol. (Trichoptera: Limnephilidae). Can. Ent. 96(6): 914-917. (2 refs., 13 figs.)

1137. Schmitz, H. 1929. Ergebnisse der Groenlandreise des "Pourquoi-pas?" 1926. Eine neue Megaselia Art (Untergattung Aphiochaeta) von Ost-Groenland. Natuurh. Maandbl. 18: 85-86. (0 ref., 2 figs.)

1138. Scholander, P.F. 1959. Supercooling and freezing in poikilotherms. 21st int. Congr. Physiol. Sci. Symposia and special lectures, pp. 77-80. (13 refs.)

1139. Scholander, P.R., W. Flagg, V. Walters, and L. Irving. 1953a. Climatic adaptation in arctic and tropical poikilotherms. Physiol. Zool. 26(1): 67-92. (63 refs., 12 figs., 9 tabs.)

1140. Scholander, P.F., W. Flagg, R.J. Hock, and L. Irving. 1953b. Studies on the physiology of frozen plants and animals in the arctic. J. cell. comp. Physiol. 42 suppl. 1. 56 pp. (103 refs., 12 figs., 10 tabs.)

1141. Schulte, G. 1975. Holarktische Artareale der Ameronothridae (Acari, Oribatei). Veroff. Inst. Meeresforsch. Bremerh. 15: 339-357. (42 refs., 3 figs., 2 tabs.)

1142. Schwarz, E.A. 1899. List of insects hitherto known from the Pribilof Islands. pp. 547-554 in D.S. Jordan, The fur seals and fur-seal islands of the North Pacific Ocean. Vol. III. Special papers relating to the fur seal and to the natural history of the Pribilof Islands. 629 pp. Govt. Printing Office, Washington. (text refs., 2 figs.)

1143. Scoresby, W., Jr. 1823. Journal of a voyage to the northern Whale-Fishery, including researches and discoveries on the eastern coast of West Greenland, made in the summer of 1822, in the ship Baffin of Liverpool. Constable, Edinburgh. 472 pp. V. Insecta (pp. 423-428) in Appendix III, pp. 416-429, List of animals met with on the Eastern coast of West Greenland with notes and memoranda. [including memoranda by Professor Jameson] (text refs.)

1144. Scudder, G.G.E. 1979. Present patterns in the fauna and flora of

Canada. pp. 87-179 in H.V. Danks (Ed.), Canada and its insect fauna. Mem. ent. Soc. Can. 108. 573 pp. (312 refs., 51 figs., 5 tabs.)

1145. Scudder, S.H., E.T. Cresson, A.R. Grote, E. Burgess, J.L. LeConte, H. Hagen, and J.H. Emerton. 1879. Insects. Hymenoptera, Nocturnal Lepidoptera, Diptera, Coleoptera, Neuroptera and Arachnida. pp. 159-161 in L. Kumlien et al., Contributions to the natural history of arctic America. Bull. U.S. natn Mus. 15. 179 pp. (0 ref.)

1146. Secher, K. 1968. [Lepidoptera in the Frederikshab district.] Grønland 1968 (8): 243-248. (In Danish) (0 ref., text figs.)

1147. Sellnick, M. 1960. Zercon hammerae, nov. spec., eine neue Milbenart aus Ost-Grönland (Acarina, Zerconidae). Ent. Meddr 29(5-6): 216-220. (1 ref., 2 figs.)

1148. Sellnick, M. and K.-H. Forsslund. 1956. Die Camisiidae Schwedens (Acar. Oribat.). Ark. Zool. 2:8(4): 473-530. (25 refs., 47 figs., keys)

1149. Serra-Tosio, B. 1972. Ecologie et biogéographie des Diamesini d'Europe (Diptera, Chironomidae). Trav. Lab. Hydrobiol. Piscic. Univ. Grenoble 63: 5-175. (236 refs., 34 figs., text tabs.)

1150. Seton, E.T. 1911. The arctic prairies; a canoe journey of 2,000 miles in search of the caribou; being the account of a voyage to the region north of Aylmer Lake. Ernest Willard, N.Y. 415 pp. [also Scribner, N.Y. 1917. (no appendixes, 308 pp.); International U. Press, N.Y., 1948.] (0 ref., 125 figs., 48 pls., 7 appendixes)

1151. Seyd, E.L. 1979. The evolution and distribution of Calyptozetes sarekensis (Acari: Oribatei). Biol. J. Linn. Soc. 12: 1-18. (69 refs., 1 fig.)

1152. Shaw, F.R. 1941. Some new species of the genus Sciara from Canada. Can. Ent. 73(9): 174-175. (0 ref., 2 figs.)

1153. Sherman, J.D., Jr. 1910. A list of Labrador Coleoptera. Jl N.Y. ent. Soc. 18(3): 173-197. (text refs.)

1154. Sherman, J.D., Jr. 1919. Dytiscidae. pp. 18-19 in Part E. Coleoptera. The Coleoptera collected by the Canadian Arctic Expedition 1913-1918. Rep. Can. Arctic Expedition 1913-1918. Vol. III: Insects. 27 pp. (0 ref.)

1155. Shewell, G.E. 1952. New Canadian blackflies (Diptera: Simuliidae). I Can. Ent. 84(2): 33-42. (3 refs., 4 figs.)

1156. Shewell, G.E. 1955. A new species of Hilara from arctic Canada (Diptera: Empidae). Can. Ent. 87(1): 45-46. (5 refs.)

1157. Shewell, G.E. 1957. Interim report on distributions of the black flies (Simuliidae) obtained in the northern insect survey. Def. Res. Bd Can. Environ. Protect. Tech. Rep. 7. 47 pp. (text refs., 47 figs.)

1158. Shewell, G.E. 1958. Classification and distribution of arctic and subarctic Simuliidae. Proc. 10th int. Congr. Ent (Montreal 1956) 1: 635-643. (12 refs., 3 figs.)

1159. Shewell, G.E. and F.J.H. Fredeen. 1958. Two new black flies from Saskatchewan (Diptera: Simuliidae). Can. Ent. 90(12): 733-738. (5 refs., 22 figs.)

1160. Shields, O. and J.C. Montgomery. 1966. The distribution and bionomics of arctic-alpine Lycaena phlaeas subspecies in North America. J. Res. Lepid. 5(4): 231-242. (28 refs., 6 figs.)

1161. Shorthouse, J.D. 1974. High arctic butterflies. Insect World Digest 1(6): 18-23. (3 refs., 10 figs.)

1162. Shorthouse, J.D. 1976. High arctic butterflies: the art of survival. Nature Canada 5(3): 42-44. (0 ref., 4 figs.)

1163. Simon, E. 1888. Arachnides recueillis au Groenland, en 1888, par M. Ch. Rabot. Bull. Soc. zool. Fr. 14: 289-291. (text refs.)

1164. Skinner, H. and L.W. Mengel. 1892. Greenland Lepidoptera. Proc. Acad. nat. Sci. Philad. 156-159. (text refs.)

1165. Skorikov, A.S. 1937. Die grönländischen Hummeln im Aspekte der Zirkumpolarfauna. Ent. Meddr 20: 37-64. (text refs., 1 fig., 1 tab.)

1166. Slack, K.V., J.W. Nauman, and L.J. Tilley. 1976. Evaluation of three collecting methods for a reconnaissance of stream benthic invertebrates. J. Res. U.S. Geol. Surv. 4(4): 491-495. (17 refs., 1 fig., 5 tabs.)

1167. Slack, K.V., J.W. Nauman, and L.J. Tilley. 1977. Benthic invertebrates in an arctic mountain stream, Brooks Range, Alaska. J. Res. U.S. Geol. Surv. 5(4): 519-527. (38 refs., 4 figs., 6 tabs.)

1168. Slack, K.V., J.W. Nauman, and L.J. Tilley. 1979. Benthic invertebrates in a north-flowing stream and a south-flowing stream, Brooks Range, Alaska. Water Resour. Bull. 15(1): 108-135. (88 refs., 5 figs., 7 tabs.)

1169. Sladen, F.W.L. 1919. Part G: Hymenoptera and Plant Galls. The wasps and bees collected by the Canadian Arctic Expedition 1913-1918. Rep. Can. Arctic Expedition 1913-1918. Vol. III: Insects. pp. 25-35. (4 figs.)

1170. Smart, R.A. 1952. Biting insects in the arctic and subarctic. J. R. Army Med. Corps 98: 8-14. (6 refs.)

1171. Smetana, A. 1965a. Staphylinini und Quediini (Col., Staphylinidae) von Newfoundland, Südost-Labrador und Nova Scotia. Acta ent. fenn. 20. 60 pp. (41 refs., 77 figs., 1 tab.)

1172. Smetana, A. 1965b. Staphylinini und Quediini von Canada und Alaska, Col. Staphylinidae. Acta Univ. lund 2(13). 18 pp. (10 refs., 9 figs.)

1173. Smetana, A. 1971. Revision of the tribe Quediini of America north of Mexico (Coleoptera: Staphylinidae). Mem. ent. Soc. Can. 79. 303 pp. (103 refs., 51 maps, 176 figs.)

1174. Smetana, A. 1978. Revision of the subfamily Sphaeridiinae of America north of Mexico (Coleoptera: Hydrophilidae). Mem. ent. Soc. Can. 105. 292 pp. (306 refs., 35 maps, 336 figs., 1 tab.)

1175. Smith, C.F. and C.S. Parron. 1978. An annotated list of Aphididae (Homoptera) of North America. Tech. Bull. N. Carol. agric. Exp. Stn 255. 428 pp. (114 refs.)

1176. Smith, C.N., W.C. McDuffie, and R.I. Sailer. 1952. Insects affecting man and animals in Alaska, faunal and control studies. Proc. 3rd Alaska Sci. Conf.: 199-200. (0 ref.)

1177. Smith, D.A. 1955. Predation on mosquitoes by a willow ptarmigan chick. Ont. fld Biol. 9: 21. (0 ref.)

1178. Smith, D.R. 1979. Nearctic sawflies IV. Allantinae: Adults and larvae (Hymenoptera: Tenthredinidae). Tech. Bull. U.S. Dep. Agric. 1595. 172 pp. (250 refs., 294 figs., keys)

1179. Smith, J.B. 1884. List of Coleoptera collected by J.R. Spencer at Fort Churchill. Appendix 5, p. 62DD in R. Bell, Observations on the geology, minerology, zoology and botany of the Labrador Coast, Hudson's Strait and Bay. Geol. Surv. Can. A. Rep. 1882-1883-1884. Pt. DD. 62 pp. (0 ref.)

1180. Smith, J.B. 1891. Contributions toward a monograph of the Noctuidae of temperate North America. Revision of the species of Mamestra. Proc. U.S. natn. Mus. 14 (851): 197-276. (text refs., 54 figs., keys)

1181. Smith, J.B. 1908. New species and genera of the lepidopterous family Noctuidae for 1907. Part II. Ann. N.Y. Acad. Sci. 18: 91-127. (text refs.)

1182. Smith, M.E. 1953. More butterflies from Alaska and the highway. Lepid. News 7: 123-126. (text ref.)

1183. Smith, S.M. 1970. The biting flies of the Baker Lake region, Northwest Territories (Diptera: Culicidae and Simuliidae). Unpub. Ph.D. thesis, Univ. Manitoba, Winnipeg. 280 pp.

1184. Smith, S.M. and R.A. Brust. 1967. Field studies of intersexes of low arctic aedine mosquitoes. Manitoba Ent. 1: 49. (7 refs.)

1185. Smith, S.M. and R.A. Brust. 1970. Autogeny and stenogamy of Aedes rempeli (Diptera, Culicidae) in arctic Canada. Can. Ent. 102 (2): 253-256. (8 refs., 1 fig.)

1186. Sommerman, K.M. 1953. Identification of Alaskan black fly larvae (Diptera, Simuliidae). Proc. ent. Soc. Wash. 55(5): 258-273. (1 ref., 35 figs., 1 tab., key)

1187. Sommerman, K.M. 1966. True-false key to species of Alaskan biting mosquitoes. Mosquito News 26(4): 540-543. (7 refs., 19 figs., key)

1188. Soper, J.D. et al. 1928. A faunal investigation of southern Baffin Island. Bull. natn Mus. Can. 53 (Biol. Ser. no. 15). 143 pp. [see McDunnough 1928]

1189. Soponis, A.R. 1977. A revision of the nearctic species of Orthocladius (Diptera, Chironomidae). Mem. ent. Soc. Can. 102. 188 pp. (139 refs., 144 figs., 16 maps, 7 + 29 tabs., keys, 2 appendixes)

1190. Sorensen, W. 1898. Arachnida Groenlandica (Acaris exceptis).

Vidensk. Meddr dansk naturh. Foren. 50: 176-235. (In Danish) (text refs.)

1191. Sotavalta, O. 1965. A revision of the genus Hyphoraia Hübner s. lat. (Lepidoptera, Arctiidae). Suom. hyönt. Aikak. 31(3): 159-197. (24 + 29 refs., 31 figs., text tabs., key)

1192. Spärck, R. 1935. Insects and Arachnids II. Diptera. The Scoresby Sound Committee's 2nd East Greenland Expedition in 1932 to King Christian IX's land. Meddr Grønland 104(16): 10-15. (8 refs.)

1193. Spärck, R. 1943. Zoogeographical remarks on the chironomid fauna of Greenland. Ent. Meddr 23(2): 432-442. (12 refs.,6 figs.)

1194. Spencer, G.J. 1966. Anoplura from British Columbia and some adjacent areas. J. ent. Soc. Br. Columb. 63: 23-30. (9 refs., 1 tab.)

1195. Spencer, K.A. 1969. The Agromyzidae of Canada and Alaska. Mem. ent. Soc. Can. 64. 311 pp. (194 refs., 552 figs., 5 tabs., keys, appendix)

1196. Staeger, C. 1845. Grønlands Antliater. Naturh. Tidsskr. (2) I: 346-369. (In Danish) (text refs.)

1197. Stage, H.H. 1943. Four days in the muskeg. Mosquito News 3(4): 127-130. (0 ref., 4 figs.)

1198. Stage, H.H. 1947. Mosquito control tests from the arctic to the tropics. Rep. Smithson. Instn 1947: 349-365. (20 refs., 1 tab.)

1199. Stage, H.H. and E.A. McKinley. 1946. A preliminary list of mosquitoes occurring in the vicinity of Nome, Alaska. Mosquito News 6(3): 131. (0 ref.)

1200. Staudinger, O. 1857a. Beitrag zur Kenntniss der nordischen Anarta-Arten. Stett. Ent. Zeitg. 18: 289-298. (text refs.)

1201. Staudinger, O. 1857b. Beitrag zur Lepidopteren-fauna Grönlands. Stett. Ent. Zeitg. 18: 299-308. (text refs.)

1202. Steel, W.O. 1958. Notes on the Omaliinae (Col., Staphylinidae). Entomologist's mon. Mag. 94: 138-142. (text refs., 24 figs.)

1203. Steel, W.O. 1964. Nomenclatorial and distributional notes on some holarctic Omaliinae (Col., Staphylinidae). Opusc. ent. 29(3): 248-250. (text refs.)

1204. Stein, P. 1910. Zur Kenntniss der Gattung Fucellia Rob. Desv. Weiner ent. Ztg. 29(1): 11-27. (text refs., key)

1205. Stephen, W.P. 1968. Biology, behaviour and orientation of bees in the arctic. Final Rep., Arct. Inst. N. Am., project ONR-383. 14 pp., figs. 1-3. (0 ref., 3 figs.)

1206. Steyskal, G.C. 1954a. The Sciomyzidae of Alaska (Diptera). Proc. ent. Soc. Wash. 56(2): 54-71. (5 + 10 refs., 12 figs., key)

1207. Steyskal, G.C. 1954b. The American species of the genus Dictya Meigen (Diptera, Sciomyzidae). Ann. ent. Soc. Am. 47(3): 511-539. (10 refs., 45 figs., key)

1208. Steyskal, G.C. 1954c. The genus Pteromicra Lioy (Diptera,

Sciomyzidae) with especial reference to the North American species. Pap. Mich. Acad. Sci. Arts Lett. 39(2): 257-269. (text refs., 9 figs., key)

1209. Steyskal, G.C. 1959. The American species of the genus Tetanocera Duméril (Diptera). Pap. Mich. Acad. Sci. Arts Lett. 44: 55-91. (text refs., 40 figs., 0 tabs., key)

1210. Steyskal, G.C., T.W. Fisher, L. Knutson, and R.E. Orth. 1978. Taxonomy of North American flies of the genus Limnia (Diptera: Sciomyzidae). Univ. Calif. Publs Ent. 83. 48 pp., 5 pls. (22 refs., 19 figs., 8 maps, 5 pls., 1 tab., keys)

1211. Stibick, J.N.L. 1976. A revision of the Hypnoidinae of the World. (Col., Elateridae). Eos, Madr. 51(1975): 143-223. (text refs., 9 charts, 4 maps, keys)

1212. Stibick, J.N.L. 1978. A revision of the Hypnoidinae of the world, Part II. (Col., Elateridae). Eos, Madr. 52(1976): 309-386. (text refs., charts 10-17, keys)

1213. Stitz, H. 1928. Die Neuropteren des arktischen Gebietes. Fauna arct. 5: 633-638. (6 refs.)

1214. Stitz, H. and W. Ramme. 1928. Nachtrag zu dem Neuropteren- und dem Orthopteren-Abschnitt. Fauna arct. 5: 855-856. (text + 1 ref.)

1215. Stocker, Z.S.J. 1972. The ecology of the streams at Char Lake, Resolute Bay, Northwest Territories. Unpub. M.Sc. thesis, Univ. Waterloo, Ontario. 122 pp.

1216. Stocker, Z.S.J. and H.B.N. Hynes. 1976. Studies on the tributaries of Char Lake, Cornwallis Island, Canada. Hydrobiologia 49(2): 97-102. (14 refs., 1 fig., 2 tabs.)

1217. Stone, A. 1938. The horseflies of the subfamily Tabaninae of the nearctic region. Misc. Publs. U.S. Dep. Agric. 305. 171 pp. (text + 24 refs., 79 figs., keys)

1218. Stone, A. 1949. A new genus of Simuliidae from Alaska (Diptera). Proc. ent. Soc. Wash. 51(6): 260-267. (1 ref., 12 figs.)

1219. Stone, A. 1952. The Simuliidae of Alaska (Diptera). Proc. ent. Soc. Wash. 54(2): 69-96. (text refs., 5 figs., keys)

1220. Stone, A., C.W. Sabrosky, W.W. Wirth, R.L. Foote, and J.R. Colson. 1965. A catalog of the Diptera of America north of Mexico. U.S. Dep. Agric.,Agric. Handbk 276. 1696 pp. (about 4800 refs., text tab.)

1221. Strand, E. 1905. Coleoptera, Hymenoptera, Lepidoptera und Araneae. Report of the second Norwegian Arctic expedition in the "Fram" 1898-1902. I(3). 30 pp. Videnskab-selsk Kristiania. (18 + text refs.)

1222. Strand, E. 1906. Die arktischen Araneae, Opiliones und Chernetes. Fauna arct. 4: 431-478. (91 refs., 3 figs., 1 tab.)

1223. Strand, E. 1914. Bemerkungen zu einigen Arbeiten über grönländische Insekten. Arch. Naturgesch. 80(A): 130-131. [A criticism of, with additions to, Nielsen, 1910] (text refs.)

1224. Strandtmann, R.W. 1967. A study of the terrestrial arthropods, especially the Acarina, of far north Alaska. Final Rep. Arct. Inst. N. Am. 9 pp. [not available; partly in Strandtmann 1971]

1225. Strandtmann, R.W. 1971. The eupodoid mites of Alaska (Acarina: Prostigmata). Pacif. Insects 13(1): 75-118. (17 refs., 15 figs., keys)

1226. Strenzke, K. von. 1952. Zur Systematik westgrönländischer Oribatiden. Zool. Anz. 149(3-4): 89-96. (14 refs.)

1227. Strenzke, K. von, H. de Lesse, and J. Denis. 1955. Microfaune du sol de l'Eqe, Groenland; vol. 1, arachnides. Paris, Hermann. 84 pp. [Expéditions polaires françaises, VII; pub. as Actualités scientifiques et industrielles, 1232] 84 pp., VIII. pls. (89 refs., 9 figs., 1 map, 8 pls.)

1228. Summers, F.M. 1957. Two mites of the genus Cheylostigmaeus, including a new species from Point Barrow, Alaska (Acarina: Stigmaeidae). Pan.-Pacif. Ent. 33(4): 163-169. (5 refs., 9 figs.)

1229. Summers, F.M. 1962. The genus Stigmaeus (Acarina: Stigmaeidae). Hilgardia 33(10): 491-537. (23 refs., 55 figs., key)

1230. Summers, R.W. 1974. The food of the snow bunting. pp. 56-58 in R.M. G. O'Brien and J.J.D. Greenwood (Eds.), Report of the University of Dundee N.E. Greenland Expedition 1972. Dundee: Dundee University N.E. Greenland Expedition. (2 refs., 2 figs.)

1231. Sutherland, P.C. 1852. Journal of a voyage in Baffin's Bay and Barrow Straits, in the years 1850-1851, performed by H.M. ships "Lady Franklin" and "Sophia", under the command of Mr. William Penny, in search of the missing crews of H.M. ships Erebus and Terror. Longman, Brown, Green and Longmans, London. 2 vols; 506 pp., 363 + ccxxxiii pp. [see White 1852]

1232. Swaine, J.M. 1919. Forest insects, including Ipidae, Cerambycidae, and Buprestidae. pp. 3-13 in Part E: Coleoptera. The Coleoptera collected by the Canadian Arctic Expedition 1913-1918. Rep. Can. Arctic Expedition 1913-1918. Vol. III: Insects. 27 pp. (text refs.)

1233. Swaine, J.M., H.C. Fall, C.W. Leng, and J.D. Sherman, Jr. 1919. Part E: Coleoptera. The Coleoptera collected by the Canadian Arctic Expedition, 1913-1918. Rep. Can. Arctic Expedition 1913-1918. Vol. III: Insects. 27 pp. [see authors]

1234. Swales, D.E. 1966. Species of insects and mites collected at Frobisher Bay, Baffin Island 1964 and Inuvik, N.W.T., 1965, with brief ecological and geographical notes. Ann. Soc. ent. Queb. 11(3): 189-199. (10 refs.)

1235. Sweatman, G.K. 1958. Redescription of Chorioptes texanus, a parasitic mite from the ears of reindeer in the Canadian arctic. Can. J. Zool. 36(4): 525-528. (6 refs., 2 figs., 1 tab.)

1236. Symposium on distribution of arctic and subarctic insects. 1958. Proc. 10th int. Congr. Ent. (Montreal 1956) 1: 613-730. [see Benson, Bruggemann, Dos Passos, Ehrlich, Freeman, Holland, Jenkins, Klots, Lindroth, Mason, Shewell, Vockeroth]

1237. Székessy, V. 1936. Revision der boreoalpinen Koleopteren auf vergleichend-anatomischer Grundlage. II. Teil. Ent. Tidskr. 57(2-3):

97-126. (38 + text refs., 20 figs.)

1238. Tambs-Lyche, H. 1967. Notes on the distribution of some arctic spiders. Astarte 28. 13 pp. (30 refs.)

1239. Tanimoto, R.M. 1971. Introductory survey of adult mosquitoes in the Yukon-Kuskokwim delta of Alaska. Mosquito News 31(4): 544-551. (7 refs., 2 figs.)

1240. Tanno, K. 1975. [Supercooling in arctic Collembola in early summer] Low Temp. Sci., B 33: 39-43. (In Japanese, English summary) (8 refs., 10 figs., 1 tab.)

1241. Tanno, K. 1976. [Permafrost and invertebrates.] pp. 127-143 in S. Kinosita (Ed.), [Joint studies on physical and biological environments in the permafrost of Alaska and northern Canada, 1974] Inst. Low Temp. Sci., Hokkaido Univ., Sapporo. (In Japanese, English summary) (45 refs., 12 tabs.)

1242. Tanno, K. and K. Shimada. 1978. Diversity in Collembola and topographic features of tundra. pp. 85-92 in [Joint studies on physical and biological environments in the permafrost of Alaska and northern Canada, 1977] Inst. Low Temp. Sci., Hokkaido Univ., Sapporo. [not seen]

1243. Taylor, P.S. 1974. Summer population and food ecology of jaegers and Snowy owls on Bathurst Island, N.W.T., emphasizing the long-tailed Jaeger. Unpub. M.Sc. thesis, Univ. Alta, Edmonton. 168 pp.

1244. Thienemann, A. 1937. Arktische Chironomidenlarven und -puppen aus dem Zoologischen Museum, Oslo. (I. Systematisches Verzeichnis. II. Metamorphose und Verbreitung von Psectrocladius barbimanus Edw.) Norsk ent. Tidsskr. 5(1): 1-7. (9 refs., 3 figs., key)

1245. Thomas, R.H. and S.F. MacLean, Jr. 1979. Community structure of soil Acari along a latitudinal transect of tundra sites in northern Alaska. Proc. 30th Alaska Sci. Conf.: 92. [abstract only]

1246. Thorell, T. 1872. [On some Arachnida from Greenland.] Öfvers. K. VetenskAkad. Förh. 29(2): 147-166. (0 ref.) (In Swedish)

1247. Thorell, T. 1878. Notice of the spiders of the "Polaris" expedition. Am. Nat.12: 393-396. (text refs.)

1248. Tilley, L.J. 1978. Some larvae of Diamesinae and Podonominae, Chironomidae from the Brooks Range, Alaska, with provisional key. Pan.-Pacif. Ent. 54(4): 241-260. (22 refs., 14 figs., key)

1249. Tilley, L.J. 1979. Some larvae of Orthocladiinae, Chironomidae from the Brooks Range, Alaska, with provisional key (Diptera). Pan-Pacif. Ent. 55(2): 127-146. (17 refs., 13 figs., key)

1250. Titus, E.G. 1911. The genera Hypera and Phytonomus (Coleoptera, family Curculionidae) in America, north of Mexico. Ann. ent. Soc. Am. 4(4): 383-473. (text refs., 11 maps, 11 pls., 1 tab., key).

1251. Tjeder, B. 1963. Three subapterous crane-flies from Alaska. (Diptera: Tipulidae). Opusc. ent. 28(3): 229-241. (16 refs., 27 figs.)

1252. Tjeder, B. 1970. Rhypholophus affinis Lundbeck, 1898, from Greenland; redescription and lectotype designation (Diptera: Tipulidae). Ent. Meddr 38(3): 253-256. (text refs., 4 figs.)

1253. Tongiorgi, P. 1966. Wolf spiders of the Pardosa monticola group (Araneae, Lycosidae). Bull. Mus. comp. Zool. Harv. 134(9): 335-359. (64 refs., 32 figs., key)

1254. Townes, H.K. 1945. The nearctic species of Tendipedini. Am. midl. Nat. 34(1): 1-206. (text refs., 261 figs., keys)

1255. Townes, H.K. 1963. Evidence among the ephialtine Ichneumonidae for a Bering migration route. pp. 153-158 in J.L. Gressitt (Ed.), Pacific Basin biogeography. (Proc. 10th Pacif. Sci. Congr. 1961). Bishop Museum Press, Honolulu. 561 pp. (2 refs.)

1256. Townes, H.K. and V.K. Gupta. 1962. Ichneumon-flies of America north of Mexico: 4. Subfamily Gelinae, tribe Hemigasterini. Mem. Am. ent. Inst. 2. 305 pp. (text refs., 116 figs., keys)

1257. Townes, H.K. and M.C. Townes. 1949. A revision of the genera of the American species of Tryphonini (Hymenoptera: Ichneumonidae), Part I, Part II. Ann. ent. Soc. Am. 42(3): 321-395; (4): 397-447. (text refs., 48 figs., keys)

1258. Townes, H.K. and M.C. Townes. 1959. Ichneumon-flies of America north of Mexico: 1. Subfamily Metopiinae. Bull. U.S. natn. Mus. 216. 318 pp. (text refs., 196 figs., keys)

1259. Townes, H.K. and M.C. Townes. 1960. Ichneumon-flies of America north of Mexico: 2. Subfamilies Ephialtinae, Xoridinae, Acaenitinae. Bull. U.S. natn. Mus. 216, pt. 2. 676 pp. (text refs., 378 figs., keys)

1260. Townes, H.K. and M.C. Townes. 1962. Ichneumon-flies of America north of Mexico: 3. Subfamily Gelinae, tribe Mesostenini. Bull. U.S. natn. Mus. 216, pt. 3. 602 pp. (text refs., 400 figs., keys)

1261. Townes, H.K. and M.C. Townes. 1978. Ichneumon-flies of America north of Mexico: 7. Subfamily Banchinae, tribes Lissonotini and Banchini. Mem. Am. ent. Inst. 26. 614 pp. (text refs., 412 figs., text maps, keys)

1262. Townsend, C.H.T. 1915. New Canadian and Alaskan Muscoidea. Can. Ent. 47(9): 285-292. (text refs.)

1263. Townsend, C.H.T. 1926. New holarctic Muscoidea. Insecutor Inscit. menstr. 14: 24-41. (text refs.)

1264. Trägårdh, I. 1901. Revision der von Thorell aus Grönland, Spitzbergen und der Bären Insel und von L. Koch aus Sibirien und Novaja Semlja beschriebenen Acariden. Zool. Anz. 25: 56-62. (text refs.)

1265. Trägårdh, I. 1904. Monographie der arktischen Acariden. Fauna arct. 4: 1-78. (23 refs., 133 figs., 1 pl., text tabs., keys)

1266. Trägårdh, I. 1917. Acari. [northeast Greenland Expedition] Meddr Grønland 43: 417-425. (13 refs., 5 figs., tabs.)

1267. Travis, B.V. 1949. Studies of mosquito and other biting insect problems in Alaska. J. econ. Ent. 42(3): 451-457. (0 ref., 1 tab.)

1268. Trelka, D.G. and B.A. Foote. 1970. Biology of slug-killing Tetanocera (Diptera: Sciomyzidae). Ann. ent. Soc. Am. 63(3): 877-895. (25 refs., 52 figs.)

1269. Trimble, R.M. 1972. Occurrence of Culiseta minnesotae and Aedes trivittatus (Diptera: Culicidae) in Manitoba including a list of mosquitoes from Manitoba. Can. Ent. 104: 1535-1537. (14 refs., 1 tab.)

1270. Tullberg, T. 1876. Collembola borealia. Nordiska Collembola. Ofvers. K. VetenskAkad. Förh. 33(5): 23-42, Pl. VIII-X. (text refs., 3 pls., 1 tab.)

1271. Turnbull, A.L., C.D. Dondale, and J.H. Redner. 1965. The spider genus Xysticus C.L. Koch (Araneae: Thomisidae) in Canada. Can. Ent. 97(12): 1233-1280. (11 refs., 179 figs., keys)

1272. Tuthill, L.D. 1943. The psyllids of America north of Mexico (Psyllidae: Homoptera). Iowa St. Coll. J. Sci. 17: 443-660. (220 refs., 313 figs., keys)

1273. Tuxen, S.L. 1934. Uber Sminthurus concolor Meinert. Zool. Anz. 106 (1-2): 4-6. (text refs., 4 figs.)

1274. Tuxen, S.L. 1952. Die Jugendstadien der nordischen Camisiiden (Acar. Orib.) und etwas über die Systematik der Erwachsenen. I. Die Gattungen Nothrus, Platynothrus und Heminothrus. Ent. Meddr 26 (5): 392-403. (text refs., text figs., keys)

1275. Twinn, C.R. 1927. Mosquitoes from Baffin Land. Can. Ent. 59(2): 47-49. (4 refs., 2 figs.)

1276. Twinn, C.R. 1935. Records of Mallophaga and other external parasites from birds at Churchill, Manitoba. Can. Ent. 67(7): 157-159. (4 refs.)

1277. Twinn, C.R. 1936. The blackflies of eastern Canada (Simuliidae, Diptera). Parts I-II. Can. J. Res. (D) 14 (9,10): 97-150. (58 refs., 15 figs., keys)

1278. Twinn, C.R. 1949. Mosquitoes and mosquito control in Canada. Mosquito News 9(2): 35-41. (5 refs., 2 tabs.)

1279. Twinn, C.R. 1950. Studies of the biology and control of biting flies in northern Canada. Arctic 3(1): 14-26. (6 refs., 11 pls.)

1280. Twinn, C.R. 1952. A review of studies of blood sucking flies in northern Canada. Can. Ent. 84(1): 22-28. (27 refs.)

1281. Twinn, C.R. 1953. Biology and control studies of biting flies. Program 1953. Science Serv., Ent. Div., Can. Dept. Agric. 45 pp. (mimeo) (64 refs., text tabs.)

1282. Twinn, C.R. 1955. Review of recent progress in mosquito studies in Canada. Mosquito News 15(4): 195-203. (52 refs.)

1283. Twinn, C.R., B. Hocking, W.C. McDuffie, and H.F. Cross. 1948. A preliminary account of the biting flies at Churchill, Manitoba. Can. J. Res. (D) 26(6): 334-357. (16 refs., 14 figs.)

1284. Ullrich, W.G. and J.M. Campbell. 1974. A revision of the apterus-group of the genus Tachinus Gravenhorst (Coleoptera: Staphylinidae). Can. Ent. 106(6): 627-644. (16 refs., 32 figs.,

1 map, 1 + 2 tabs., key)

1285. Ulmer, G. 1932. Die Trichopteren, Ephemeropteren und Plecopteren des arktischen Gebietes. Fauna arct. 6: 207-226. (74 refs.)

1286. United States Polar Expedition 1881-1883. 1885. Rept. International Polar Exped. to Pt. Barrow, Alaska. U.S. Govt. printing office, Washington. [see Murdoch and Riley 1885]

1287. U.S. Tundra Biome open literature and dissertations. 1977. C.R.R.E. L., Hanover, N.H. 22 pp.

1288. Usinger, R.L. 1960. Observations on the biology of Chiloxanthus stellatus (Curtis) and C. arcticus (Sahlberg). Pan.-Pacif. Ent. 36(10):189-190. (4 refs.)

1289. Van Achterberg, G. 1975. A revision of the tribus Blacini (Hymenoptera, Braconidae, Helconinae). Tijdschr. Ent. 118(7): 159-322. (30 refs., 476 figs., keys)

1290. Van Duzee, E.P. 1919. Part F. Hemiptera. Hemiptera of the Canadian Arctic Expedition. Rep. Can. Arctic Expedition 1913-1918. Vol. III: Insects. 5 pp. (0 ref.)

1291. Van Duzee, E.P. 1921. Orthoptera, Neuroptera, Hemiptera and Lepidoptera from the Pribilof Islands, Alaska. pp. 193-195 in G.D. Hanna et al., Insects of the Pribilof Islands, Alaska. Proc. Calif. Acad. Sci. ser. 4, 11(14): 153-195. (0 ref.)

1292. Van Duzee, M.C. 1921. A new species of the dipterous family Dolichopodidae from the Pribilof Islands, Alaska. pp. 167-168 in G.D. Hanna et al., Insects of the Pribilof Islands, Alaska. Proc. Calif. Acad. Sci. ser 4, 11(14): 153-195. (0 ref.)

1293. Van Duzee, M.C., F.R. Cole, and J.M. Aldrich. 1921. The dipterous genus Dolichopus Latreille in North America. Bull. U.S. natn. Mus. 116. 304 pp., 16 pls. (20 + text refs., 217 figs., key)

1294. Van Dyke, E.C. 1921. Coleoptera from the Pribilof Islands, Alaska. pp. 156-166 in G.D. Hanna et al., Insects of the Pribilof Islands, Alaska. Proc. Calif. Acad. Sci. ser. 4, 11(14): 153-195. (text refs.)

1295. Van Dyke, E.C. 1928. The species of the genus Lepyrus Germ. (Coleoptera - Curculionidae) in North America. Pan.-Pacif. Ent. 5(2): 53-58. (text refs., key)

1296. Van Dyke, E.C. 1938. A review of the genus Chrysolina Motschulsky in North America (Coleoptera-Chrysomelidae). Bull. Brooklyn ent. Soc. 33(2): 45-58. (text refs., key)

1297. Van Helsdingen, P.J. 1974. The affinities of Wubana and Allomengea with some notes on the latter genus (Araneae, Linyphiidae). Zool. Medded. 46(22): 295-321. (33 refs., 25 figs., text tabs.)

1298. Vibe, C. 1950. Some insects new to Greenland. Ent. Meddr 25(6): 419-420. (0 ref.)

1299. Vibe, C. 1953. [Zoogeographic problems in Greenland.] Grønland 1953(11): 432-436. (In Danish) (0 ref., 2 figs.)

1300. Vibe, C. 1956. The zoogeography of Greenland. Proc. 14th int. Congr. Zool. (Copenhagen 1953): 126-127. (0 ref.)

1301. Vickery, V.R. 1967. The Orthoptera of Alaska, Yukon and the Mackenzie district of the northwest territories. Trans. Am. ent. Soc. 93(3): 249-278. (46 refs., 42 figs.)

1302. Vickery, V.R., D.E. Johnstone, and D.K. McE. Kevan. 1974. The orthopteroid insects of Quebec and the Atlantic Provinces of Canada. Mem. Lyman ent. Mus. Res. Lab. 1. i + ii + 204 pp. (117 refs., 130 figs., 53 maps, keys)

1303. Viereck. H.L. 1923. Hymenoptera. pp. 229-236 in W.L. McAtee (Ed.), A biological survey of the Pribilof Islands, Alaska. Part II, Insects, arachnids and chilopods of the Pribilof Islands, Alaska. N. Am. Fauna 46: 129-255. (text refs.)

1304. Viets, K. 1931. Hydracarinen der Fauna Arctica. Fauna arct. 6: 1-8. (25 refs., 2 tabs.)

1305. Vockeroth, J.R. 1952a. The specific status of Aedes pionips Dyar (Diptera: Culicidae). Can. Ent. 84(8): 243-247. (11 refs., 8 figs.)

1306. Vockeroth, J.R. 1952b. A new nearctic species of Rhaphium, with notes on other species (Diptera: Dolichopodidae). Can. Ent. 84 (9): 276-280. (1 ref., 2 figs., key)

1307. Vockeroth, J.R. 1954a. Notes on northern species of Aedes, with descriptions of two new species (Diptera: Culicidae). Can. Ent. 86(3): 109-116. (5 refs., 6 figs.)

1308. Vockeroth, J.R. 1954b. Notes on the identities and distributions of Aedes species of northern Canada, with a key to the females (Diptera: Culicidae). Can. Ent. 86(6): 241-255. (17 refs., 16 figs., 1 tab., key)

1309. Vockeroth, J.R. 1958. Distribution patterns of the Scatomyzinae (Diptera, Muscidae). Proc. 10th int. Congr. Ent. (Montreal 1956) 1: 619-625. (4 refs., 6 figs., 1 tab.)

1310. Vockeroth, J.R. 1972. A new nearctic genus of Mycetophilidae (Diptera) with a stenopterous female. Can. Ent. 104(10): 1529-1533. (0 refs., 11 figs.)

1311. Wagner, R.J., C. Dejong, M.K. Leung, J. McLintock, and J.O. Iversen. 1975. Isolations of California encephalitis virus from tundra mosquitoes. Can. J. Microbiol. 21(4): 574-576. (10 refs., 2 tabs.)

1312. Wahlgren, E. 1900. Collembola während der schwedischen Grönlands-expedition 1899 auf Jan Mayen und Ost-Grönland eingesammelt. Ofvers. K. VetenskAkad. Förh. 57(3): 353-375. (text refs., 8 figs., 1 tab.)

1313. Wahlgren, E. 1907. Collembola. Report of the Second Norwegian Arctic Expedition in the Fram, 1898-1902. II(10). 6 pp. Videnskab-Selskabet, Kristiania. (text refs., 4 figs.)

1314. Wahlgren, E. 1919. [West arctic element in Scandinavian alpine fauna.] Ent. Tidskr. 40(1): 22-32. (In Swedish, German summary) (text refs.)

1315. Walker, D. 1860. Notes on the zoology of the last arctic expedition under Captain Sir F.L. M'Clintock. J. R. Dublin Soc. 3: 61-77. (0 ref., text tabs.)

1316. Walker, E.M. 1920. Part J: Orthoptera. Orthoptera collected in the Canadian arctic. Rep. Can. Arctic Expedition 1913-1918. Vol. III: Insects. 4 pp. (text refs.)

1317. Walker, E.M. 1941. New records of Odonata from Manitoba. Can. Ent. 73(2): 35-36. (0 ref.)

1318. Walker, E.M. 1943. The subarctic Odonata of North America. Can. Ent. 75(5): 79-90. (32 refs., 9 figs.)

1319. Walker, E.M. 1947. Further notes on the subarctic Odonata of North America. Can. Ent. 79(4): 62-67. (text refs.)

1320. Walker, E.M. 1951. The Odonata of the Northern Insect Survey. Can. Ent. 83(10): 269-278. (5 refs.)

1321. Walker, E.M. 1953. The Odonata of Canada and Alaska, volume 1, part 1: general; part 2: the Zygoptera-Damselflies. Univ. Toronto Press. 292 pp. (192 refs., 44 pls., keys)

1322. Walker, E.M. 1958. The Odonata of Canada and Alaska. Volume II. Part III. The Anisoptera - four families. Univ. Toronto Press. 318 pp. (92 refs., 64 pls., keys)

1323. Walker, E.M. and P.S. Corbet. 1975. The Odonata of Canada and Alaska. Volume III. Part III: The Anisoptera - three families. Univ. Toronto Press. 307 pp. (288 refs., 45 pls., keys)

1324. Walkley, L.M. 1960. The genus Delomerista. pp. 362-373 in H.K. Townes and M.C. Townes. 1960. Ichneumon-flies of America north of Mexico: 2. Subfamilies Ephialtinae, Xoridinae, Acaenitinae. Bull. U.S. natn. Mus. 216, pt. 2. 676 pp.

1325. Walley, G.S. 1931. Corixidae from the environs of Hudson Bay (Hemip., Corixidae). Can. Ent. 63(10): 238-239. (text ref., 4 figs.)

1326. Walley, G.S. 1936. A report on some Braconidae from Baffin Island (Hymenoptera). Can. Ent. 68(8): 187-190. (0 ref.)

1327. Walley, G.S. 1940. A synopsis of the North American species of Syndipnus (Hymen., Ichneumonidae). Can. Ent. 72(7): 135-145. (6 refs., 8 figs.)

1328. Walley, G.S. 1941. Some new and little known Canadian Ichneumonidae (Hymenoptera). Can. Ent. 73(9): 164-170. (text refs.)

1329. Walley, G.S. 1960. The nearctic species of Scambus Hartig. pp. 14-79 in H.K. Townes and M.C. Townes. 1960. Ichneumon-flies of America north of Mexico: 2. Subfamilies Ephialtinae, Xoridinae, Acaenitinae. Bull. U.S. natn. Mus. 216, pt. 2. 676 pp.

1330. Walley, G.S. 1967. Nearctic species of the Diadegma stenosonus complex (Hymenoptera: Ichneumonidae). Can. Ent. 99(9): 925-943. (10 refs., 16 figs., key)

1331. Wallis, J.B. 1939. The genus Ilybius Er. in North America (Coleoptera, Dytiscidae). Can. Ent. 71(9): 192-199. (text refs., key)

1332. Walton, W.R. 1923. Family Calliphoridae. pp. 228 in W.L. McAtee (Ed.), A biological survey of the Pribilof Islands, Alaska. Part II, Insects, arachnids and chilopods of the Pribilof

Islands, Alaska. N. Am. Fauna 46: 129-255. (text refs.)

1333. Washburn, R.H. 1963. Distribution of bumblebees (Bombus) and their social parasites (Psithyrus) in Alaska. Proc. 14th Alaska Sci. Conf.: 29-37. (4 refs., 2 tabs.)

1334. Washburn, R.H. 1966. Honey production in Alaska. Proc. 17th Alaska Sci. Conf.: 77. [abstract only]

1335. Watson, D.G., J.J. Davis, and W.C. Hanson. 1966. Terrestrial invertebrates. pp. 565-584 in N.J. Wilimovsky and J.N. Wolfe (Eds.), Environment of the Cape Thompson Region, Alaska. U.S. Atomic Energy Comm. [Clearinghouse for Federal Sci. and Tech. Inf., Springfield, Va., PNE-481] 1250 pp. (5 refs., 6 tabs)

1336. Watson, W.Y. 1954. Two new species of Coccinellidae (Coleoptera). Can. Ent. 86(1): 45-47. (text refs., 9 figs.)

1337. Watt, W.B. 1968. Adaptive significance of pigment polymorphisms in Colias butterflies. I. Variation in melanin pigments in relation to thermoregulation. Evolution 22(3): 437-458. (33 refs., 11 figs., 6 tabs.)

1338. Webb, J.E. and H.M. Graham. 1956. Observations on some filth flies in the vicinity of Fort Churchill, Manitoba, Canada, 1953-54. J. econ. Ent. 49(5): 595-600. (1 ref., 3 tabs.)

1339. Weber, N.A. 1948. Opportunities for entomological research in the arctic. Ent. News 59(10): 253-257. (0 ref.)

1340. Weber, N.A. 1949. Late summer invertebrates, mostly insect, of the Alaskan arctic slope. Ent. News 60(5): 118-128. (0 ref.)

1341. Weber, N.A. 1950. A survey of the insects and related arthropods of arctic Alaska. Trans. Am. ent. Soc. 76(6): 147-206. (51 refs., 10 pls.)

1342. Weber, N.A. 1953. Arctic Alaskan Hymenoptera and Coleoptera. Ent. News 64(10): 256-260. (2 refs.)

1343. Weber, N.A. 1954. Arctic Alaskan Diptera. Proc. ent. Soc. Wash. 56 (2): 86-91. (0 ref.)

1344. Weber, N.A. 1970. Extension of insect ranges on the arctic Alaskan slope. Ent. News 81(5): 128. (text refs.)

1345. Webster, F.M. 1894. Butterflies common to Norway and arctic North America. Can. Ent. 26(5): 117-120. (text ref.)

1346. Welch, H.E. 1960a. Hydromermis churchillensis n. sp. (Nematoda: Mermithidae) a parasite of Aedes communis (DeG.) from Churchill, Manitoba, with observations of its incidence and bionomics. Can. J. Zool. 38(3): 465-474. (22 refs., 11 figs., 1 tab.)

1347. Welch, H.E. 1960b. Effects of protozoan parasites and commensals on larvae of the mosquito Aedes communis (Degeer) (Diptera: Culicidae)at Churchill, Manitoba. J. Insect Pathol. 2(4): 386-395. (17 refs., 3 figs., 2 tabs.)

1348. Welch, H.E. 1972. Chironomid energetics. pp. 41-47 in F.H. Rigler (Ed.), A. Rep. 1971-1972 Char Lake Project PF-2. Dept. Zool. Univ. Toronto. 85 pp. [largely in Welch 1973, 1976]

1349. Welch, H.E. 1973. Emergence of Chironomidae (Diptera) from Char Lake, Resolute, Northwest Territories. Can. J. Zool. 51(11): 1113-1123.

1350. Welch, H.E. 1976. Ecology of Chironomidae (Diptera) in a polar lake. J. Fish. Res. Bd Can. 33(2): 227-247. (25 refs., 23 figs., 12 tabs.)

1351. West, A.S. 1951. The Canadian arctic and subarctic mosquito problem. Proc. New Jers. Mosq. Exterm. Assoc. 38: 105-110. (0 ref.)

1352. West, A.S. and D.W. Jenkins. 1951. Plant feeding habits of northern mosquitoes studied with radioisotopes. Mosquito News 11(4): 217-219. (3 refs.)

1353. Wheeler, M.R. and L.H. Throckmorton. 1960. Notes on Alaskan Drosophilidae (Diptera), with the description of a new species. Bull. Brooklyn ent. Soc. 50(3): 134-143. (8 refs., 2 pls.)

1354. Wheeler, W.M. 1917. The mountain ants of western North America. Proc. Am. Acad. Arts. Sci. 52(8): 455-569. (69 refs., 1 fig., text tabs.)

1355. White, A. 1851. Notice and list of insects. Vol. II, pp. 354-363 in J. Richardson. Arctic searching expedition; a journal of a boat voyage through Rupert's Land and the arctic sea in search of the Discovery ships under command of Sir John Franklin. Longman, Brown, Green and Longman's, London. 426 pp. (0 ref., 1 tab.)

1356. White, A. 1852. Insects and Aptera. Vol. II, pp. ccviii-ccxi in P.C. Sutherland, Journal of a voyage in Baffin's Bay and Barrow Straits, in the years 1850-1851, performed by H.M. ships "Lady Franklin" and "Sophia", under the command of Mr. William Penny, in search of the missing crews of H.M. ships Erebus and Terror. Longman, Brown, Green and Longman's,London. 363 pp + ccxxiii pp., appendixes. (text refs., 8 figs.)

1357. Whitfield, D.W.A. 1977. Energy budgets and ecological efficiencies on Truelove lowland. pp. 607-620 in L.C. Bliss (Ed.), Truelove lowland, Devon Island, Canada: A high arctic ecosystem. Univ. Alta. Press, Edmonton. 714 pp. (15 refs., 5 figs., 4 tabs.)

1358. Whittaker, J.B. 1974. Interactions between fauna and microflora at tundra sites. pp. 183-196 in A.J. Holding, O.W. Heal, S.F. MacLean, Jr., and P.W. Flanagan (Eds.), Soil organisms and decomposition in tundra. Tundra Biome Steering Committee, Stockholm, Sweden. 398 pp. (15 refs., 6 figs., 4 tabs.)

1359. Wickham, H.F. 1923. Coleoptera. pp. 150-157 in W.L. McAtee (Ed.), A biological survey of the Pribilof Islands, Alaska. Part II, Insects, arachnids and chilopods of the Pribilof Islands, Alaska. N. Am. Fauna 46: 129-255. (text refs.)

1360. Williams, R.C. 1931. On some northern Lepidoptera Rhopalocera. Ent. News 42(6): 157. (text ref.)

1361. Wilson, C.S. 1951. Control of Alaskan biting insects. Publ. Hlth Rep. 66(29): 917-922. (5 refs.)

1362. Wilson, C.S. 1953. Mosquito control. Publ. Hlth Rep. 68(5): 536-537. (0 ref.)

1363. Wirth, W.W. 1971. The brine flies of the genus Ephydra in North America (Diptera: Ephydridae). Ann. ent. Soc. Am. 64(2): 357-

377. (64 refs., 41 figs.)

1364. Wirth, W.W. and F.S. Blanton. 1969. North American Culicoides of the pulicaris group (Diptera: Ceratopogonidae). Fla Ent. 52(4): 207-243. (14 refs., 15 figs., 1 tab., key)

1365. Wolff, N.L. 1950. Crambus bonifatellus Hulst, an unexpected component of the Lepidopterous fauna of Greenland. Ent. Meddr 25(6): 421-424. (0 ref., 4 figs.)

1366. Wolff, N.L. 1964. The Lepidoptera of Greenland. Meddr Grønland 159(1). 74 pp. (55 refs., 55 figs., 21 pls.)

1367. Wolff, N.L. 1971. [Additions and corrections to the list of the Lepidoptera of Greenland.] Ent. Meddr 39(1): 71-79. (In Danish) (7 refs., 5 figs.)

1368. Wolff, N.L. 1975. [The Nomophila spp. (Lepidoptera, Pyralidae) found in the north Atlantic area.] Ent. Meddr 43(3): 129-135. (In Danish) (6 refs., 12 figs.)

1369. Wong, H.R. 1968. Pristiphora gelida, a new species from Alaska (Hymenoptera: Tenthredinidae). J. nat. Hist. 2(2): 185-186. (0 ref., 4 figs.)

1370. Wong, K.K. and D.M. McLean. 1978. Canadian bunyavirus morphogenesis in arctic mosquitoes. pp. 380-381 in J.M. Sturgess (Ed.), Electron microscopy 1978, Vol. II. Biology. (9th int. Congr.) XXXII + 718 pp. Microscopical Society of Canada. (0 ref., 3 figs.)

1371. Wood, D.M. 1977. Notes on the identities of some nearctic Aedes mosquitoes. Mosquito News 37(1): 71-81. (43 refs., 12 figs.)

1372. Wood, D.M. 1978. Taxonomy of the nearctic species of Twinnia and Gymnopais (Diptera: Simuliidae) and a discussion of the ancestry of the Simuliidae. Can. Ent. 110(12): 1297-1338. (32 refs., 64 figs., keys)

1373. Wood, D.M., P.T. Dang, and R.A. Ellis. 1979. The mosquitoes of Canada (Diptera: Culicidae). The insects and arachnids of Canada, Part 6. Agric. Can. Publ. 1686. 390 pp. (500 refs., 238 figs., 75 maps, 75 pls., 3 tabs., keys)

1374. Woodley, R.E. 1969. Boloria polaris (Nymphalidae) in northern British Columbia. J. Lepid. Soc. 23(2): 113-115. (3 refs., 1 fig.)

1375. World Health Organization. 1962. Environmental Biology Unit. Insects of medical and veterinary importance in high latitudes, with particular reference to North America. WHO Conference on medicine and public health in the arctic and antarctic. Arctic Conference document no. 28. 9 pp. (17 + 13 refs.)

1376. Wyatt, C. 1957. Collecting on the Mackenzie and in the western arctic. Lepid. News 11(1-3): 47-53. (0 ref.)

1377. Yarrow, I.H.H. 1955. Bombus (Alpinobombus) hyperboreus clydensis n. ssp., from Baffin Island, North-west Territories, Canada. Ann. Mag. nat. Hist. 12(8): 151-152. (6 refs.)

1378. Yoshimoto, C.M. 1971. Revision of the genus Euderus of America north of Mexico. Can. Ent. 103(4): 541-578. (18 refs., 60 figs., 2 tabs., keys)

1379. Yoshimoto, C.M. 1973. Revision of the genus *Chrysocharis* Förster (subgenus *Chrysocharis* s. str.) (Eulophidae: Chalcidoidea) of America north of Mexico. *Can. Ent.* 105(11): 1377-1405. (27 refs., 60 figs., 1 tab., key)

1380. Young, F.N. 1960. Notes on the water beetles of Southampton Island in the Canadian arctic (Coleoptera: Dytiscidae and Haliplidae). *Can. Ent.* 92(4): 275-278. (8 refs.)

1381. Zunker, M. 1932. Die Mallophagen des arktischen Gebietes. *Fauna arct.* 6: 281-294. (12 refs., 14 figs., key)

1382. Zweite deutsche Nordpolarfahrt. 1874. Geographische Gesellschaft in Bremen. Die zweite Deutsche Nordpolarfahrt in den Jahren 1869 und 1870. Bd. 2. [See Homeyer, Koch, Gerstacker and Pansch]

INDEX TO SUBSIDIARY AUTHORS

GEOGRAPHIC INDEX

Areas covered include Greenland and North America north of tree line, with exceptions noted in the introduction.

Citations are indexed only under the most specific entry. Multiple locations within an area, or multiple areas within a country, are not indexed individually.

Entries that are not underlined contain only taxonomic information (taxonomy, description, collecting records, etc.). Underlined entries include also ecological or zoogeographical information and discussion.

TAXONOMIC INDEX

The taxonomic arrangement of classes and orders follows that adopted in Danks 1979 (Danks, H.V. (Ed.). 1979. Canada and its insect fauna. Mem. ent. Soc. Can. 108. 573 pp.); families are arranged alphabetically within orders.

Lower taxa (e.g. families within an order, or orders within a class) are not indexed separately when numerous such taxa are treated in the same publication. Indexing of families in the Collembola and in small orders of insects has been limited.

Entries that are not underlined contain only taxonomic information (taxonomy, descriptions, collecting records, checklists, distribution records, identification, synonymy, phylogeny). Underlined entries include also ecological or zoogeographic information and discussion.

For general treatments see also Composition of fauna in the subject index.

SIPHONAPTERA 33, 136, 396, 445, 464, 506, 507, 570, 571, 572, 573, 574, 575, 576, 578, 579, 603, 604, 645, 646, 657, 897, 898, 1044, 1276

LEPIDOPTERA 30, 31, 41, 55, 189, 287, 295, 312, 314, 336, 387, 400, 402, 403, 415, 417, 420, 443, 451, 461, 521, 544, 597, 599, 655, 677, 699, 704, 730, 761, 773, 790, 865, 925, 938, 947, 1010, 1016, 1019, 1020, 1021, 1022, 1023, 1024, 1025, 1026, 1146, 1201, 1221, 1291, 1314, 1315, 1366, 1367

Arctiidae 318, 389, 829, 908, 1035, 1191

Gelechiidae 900

Geometridae 246, 265, 366, 396, 399, 695, 707, 715, 786, 873, 874, 876, 878, 879, 880, 932, 936, 1035, 1047, 1053, 1093, 1164

Lymantriidae 86, 365, 396, 443, 707, 783, 987, 1108, 1139, 1145, 1164

Noctuidae 50, 86, 87, 338, 388, 399, 502, 503, 707, 716, 783, 791, 867, 875, 877, 1035, 1047, 1048, 1053, 1145, 1164, 1180, 1181, 1200, 1360

Pterophoridae 48, 51, 728, 872

Pyralidae 48, 366, 399, 455, 522, 706, 901, 933, 935, 939, 941, 942, 943, 944, 1047, 1365, 1368

Sphingidae 829

Tortricidae 48, 366, 396, 398, 399, 715, 792, 871, 974, 975, 1046, 1048

Papilionoidea (and Hesperiidae) 86, 140, 194, 216, 288, 292, 309, 330, 331, 333, 342, 343, 370, 389, 399, 405, 412, 419, 448, 473, 581, 615, 687, 702, 703, 707, 729, 744, 798, 829, 830, 906, 907, 908, 1017, 1018, 1022, 1023, 1024, 1025, 1026, 1161, 1162, 1182, 1345, 1360, 1376

Lycaenidae 369, 397, 1008, 1160

Nymphalidae 174, 182, 256, 332, 341, 368, 396, 701, 703, 783, 866, 1008, 1035, 1047, 1048, 1053, 1093, 1143, 1164, 1374

Pieridae 321, 408, 610, 611, 612, 613, 614, 700, 829, 866, 931, 1008, 1035, 1047, 1048, 1053, 1083, 1143, 1164, 1337

Satyridae 231, 300, 301, 302, 303, 335, 337, 783, 831, 869, 1008

TRICHOPTERA 42, 44, 94, 237, 390, 480, 508, 598, 662, 677, 692, 726, 740, 785, 865, 885, 915, 922, 923, 924, 927, 948, 968, 1029, 1037, 1073, 1090, 1091, 1092, 1133, 1134, 1135, 1136, 1145, 1285, 1291

HYMENOPTERA 30, 55, 423, 544, 597, 601, 677, 681, 720, 777, 829, 836, 837, 1010, 1342

Apidae 86, 189, 293, 393, 395, 396, 423, 424, 425, 426, 427, 438, 563, 568, 623, 695, 782, 790, 829, 865, 881, 911, 912, 913, 916, 925, 930, 947, 1000, 1001, 1052, 1053, 1055, 1056, 1057, 1145, 1165, 1169, 1205, 1221, 1303, 1315, 1333, 1334, 1377

SUBJECT INDEX

This index excludes entries of a purely taxonomic nature that are treated in the geographic and taxonomic indexes (entries there that are not underlined: taxonomy, descriptions, collecting records, etc.)